Challenges Between Competition and Collaboration

Springer
Berlin
Heidelberg
New York
Hong Kong
London
Milan
Paris
Tokyo

Peter Sachsenmeier
Martin Schottenloher
(Editors)

Challenges Between Competition and Collaboration

The Future of the European Manufacturing Industry

With 107 Figures

Springer ewf europe

Professor Peter Sachsenmeier
IMAG Information Management AG
Blumenaustraße 28
8200 Schaffhausen
Switzerland
peter_sachsenmeier@imag-web.com

Professor Martin Schottenloher
Universität München
Mathematisches Institut
Theresienstraße 39
80333 München
Germany
martin.schottenloher@mathematik.uni-muenchen.de

ISBN 3-540-40169-5 Springer-Verlag Berlin Heidelberg New York

Cataloging-in-Publication Data applied for
A catalog record for this book is available from the Library of Congress.
Bibliographic information published by Die Deutsche Bibliothek
Die Deutsche Bibliothek lists this publication in the Deutsche Nationalbibliografie; detailed bibliographic data is available in the Internet at <http://dnb.ddb.de>.

Springer-Verlag Berlin Heidelberg New York
a member of BertelsmannSpringer Science+Business Media GmbH

http://www.springer.de

Printed in Germany

Hardcover-Design: Erich Kirchner, Heidelberg

SPIN 10932360 42/3130-5 4 3 2 1 0 – Printed on acid-free paper

Contents

Plenary Presentations

Discussion Forums

Invited Contributions

Ewf Information

Conference Information

Introduction

Peter Sachsenmeier, Martin Schottenloher

Editors

Context

The European manufacturing landscape is changing fast, and companies all over Europe are scrambling to reposition themselves in the competitive arena.

Amid all the emerging alliances, networks and relationships, competition and collaboration need to be redefined.

ewf europe is a European industrial association actively involved in redefining industrial practices since 1998. It devoted its second bi-annual conference e.e.u.c. in Brussels at the end of 2002 to "Challenges between Competition and Collaboration. The Future of European Manufacturing Industry".

We had identified that many industries are currently on the threshold of new forms of co-operation. The industrial landscape is full of alliances, strategic partnerships, collaborative networks and other attempts at managing increasingly complex partnerships. The aim of all of these is to achieve competitive advantage, shorter development cycles, and innovation.

In the new networks, firms encounter each other both as competitors and as partners. This demands new management qualities, as well as new forms of behaviour and rules in dealing with each other.

From our study of the competitive landscape in Europe, we postulated that the new challenges would only be met successfully by those companies which focus on their core competences and manage to translate these into added value and profits. There is, so we wrote in the invitation, a need for companies to change their processes, dramatically. Beyond company borders.

How do companies manage the split between competition and collaboration, in practice? How are companies positioning themselves in the face of the new challenges? What are the strategies for survival? What would help companies to make better decisions vis-à-vis the challenges in the new collaborative networks? What experiences exist? These are some of the questions we put to contributors and participants alike, and we hope the questions we put to contributors and participants alike, and we hope that you will find them amply discussed in this volume.

Our idea to create Networks of Excellence as a special outcome of the conference attracted a great deal of advance attention. The idea was to spearhead theory and practice of these emerging forms of intensive collaboration. We put ourselves forward as the courageous thinkers needed to flesh out the idea, put it into practice, establish a programme of activities to promote the idea of such networks of excellence, establish rules and procedures together with all those interested in this new form of industrial collaboration. Our intent was to help participating firms and individuals to address issues between competition and collaboration across industries and beyond geographical, linguistic and cultural borders.

Important Results

The conference attracted considerable attention. In the planning of the conference, the topic "Network of Excellence" became ever more dominant, and the establishment of possible networks among the participants one of the prime intentions of the conference.

The presentations showed surprising parallels between various industrial segments, be they automotive, aerospace, electrical engineering, or machine manufacturing, and also, similar problems, as well as similar approaches to solutions. As ewf europe President Herbert Köpplinger says, »…the holistic ewf approach of the Networks of Excellence for the overall manufacturing industry makes absolute sense. «

Networking, and in particular networks of excellence, was found to be a very topical theme. Many questions were raised, few definitive answers found. It appears that we are at the stage where we can pose the right questions and identify the direction in which we want to go. As one speaker, Peter Köpf, put it, »We are in the thick of things, and yet we are only at the beginning. «

Prior to the conference, we had intuitively perceived the need to tackle a number of problems, and the conference reinforced this view. The majority of these relate to capabilities, skills, methods and procedures for new forms of collaboration among companies. There was a unilateral feeling among participants that the new forms of partnership are all related to the theme of "Networks of Excellence".

People and organisational issues, such as trust and confidence among those participating in the new networks, were mentioned many times as prerequisites for the success of any networking initiatives. We were all agreed on the need for new forms of co-operation, and on the need for the establishment of a framework of rules, methods and procedures. It was also acknowledged that cultural factors and organisational values play an important role.

Technology took second place with regard to these new forms even though many collaborative (support) technologies were mentioned, and quite a few shown at the exhibition accompanying the conference.

Important issues raised in connection with networks included, among others, contractual difficulties, finances and capital deployment, risk and benefit sharing, human resources, intellectual property rights, and information flows.

From the discussions, it became clear that more joint research is needed in order to be able to implement the desired new forms of partnership and collaboration.

As a consequence, several proposals were put forward to put in place concrete Networks of Excellence among participants. Wilhelm Becker, Peter Stehle, and Peter Köpf were leading voices among many to call for the establishment of a Network of Excellence among members of the automotive community, now called the Network of Automotive Excellence (NoAE).

This Volume

This book contains the proceedings of the conference It contains all the presentations, most of them written up as articles, the results of the discussion forums, and a number of complementary articles which the editors have solicited. We have organised the book in accordance with these three parts.

Plenary Presentations

The presentations in plenary resulted in ten articles.

In his opening address, Hans-Jörg Bullinger and his co-author Dietmar Fischer set out the theme of the conference (apparent in the title). In their contribution, they insisted that »those who act with suppliers, customers and partners in global competence and co-operation networks and decisively use the advantages of innovative technologies and products will be successful.«

Hervé Péro looks at the changes in industries in Europe, and he describes the EU priority areas for industrial collaboration. Interestingly, he emphasizes that the shift in industrial innovation is more and more towards societal issues. He calls for a co-ordinated effort in research in order to ensure synergies between EU and national actions, and exhorts us to concentrate our efforts towards sustainable development and knowledge-based enterprises.

Ludolf Plass vividly sets out the issues connected with plant engineering on a global scale. The challenge in his highly complex industry is to achieve «the latest state of technology at the lowest price in the shortest time… turnkey at a fixed price». He points out that, despite many superficial differences, the similarities and common points among different industries are surprising, and can be exploited for benchmarking.

Steve Nevey of Jaguar Racing takes us into a world where competition is the name of the game. He provides substantial insights into how to succeed in the business of designing and manufacturing a high performance, precision product. One of his main tenets is that communication with all aspects of the enterprise and the customer needs to be encouraged at all times. Full commitment of everyone in the organisation is important to effectively manage and implement collaboration. He ends by telling us that Jaguar Racing WILL win the Formula One World championship.

Hans-Erich Mundt introduces into long term partnering, in the aerospace industries, where life cycles can be longer than thirty years. The key elements of his company's sourcing strategy include procurement marketing, procurement policy, supplier evaluation and development, and joint sourcing. In his industry, it is important that industrial partners show a long term willingness to cooperate and share the risks and opportunities.

Wilhelm Becker analyses the automotive markets in substantial detail and sets out the new tasks for automobile manufacturers, and the new demands which they place on their network because of these changes. His thesis is that a proper re-think of the cooperation processes leads to the establishment of networks. He sets out the case for a network of automotive excellence, and describes the different roles and tasks to be fulfilled, as well as fundamental changes involved. He challenges us to start an initiative as partners: »Future tasks in the automobile industry can only be solved jointly. A "Network of Automotive Excellence" cannot be created overnight. [Therefore] Manufacturers, suppliers and service providers but also trade associations and other organisations are called upon to work jointly in a spirit of dialogue for the future: setting up a "Network of Automotive Excellence".«

The second speaker from the automotive industry, Peter Stehle, contrasts the situations in which OEMs and suppliers find themselves. He describes ways in which European automobile suppliers can preserve and improve their competitive abilities, such as simplification of processes, focus on long-term partnerships, focus on core competences, and permanent efficiency improvements. To him, »becoming quicker, better and cheaper is in the long run, not going to be possible by going alone…«: another call to start a NoAE initiative as partners!

Peter Köpf, as the third automotive contributor, looks at the future role of tier 1 enterprises. According to him, collaboration extends to organisational, people and technology issues, including an alignment of IT. Standardisation continues to be a real need. Emerging imbalances in financing development are a particular concern, and require new business models.

Erastos Filos looks at the challenges faced by Europe in its attempt to create the largest knowledge-based economy in the world. These consist in the definition of value, issues of trust and relationships, the importance of knowledge and internetworking. He outlines the EU's Lisbon strategy, and the objectives of eEurope 2005. His contribution includes a detailed presentation of the EU's R&D activities to address business and work challenges, together with project examples. Of particular importance is his outline of the R&D activities and priorities foreseen under the 6th Framework Programme.

Peter Sachsenmeier puts the concept *Networks of Excellence* (NoE) to three tests, namely those of relevance, feasibility and sustainability. He defines and explores the elements of the term, and proposes to take a look at the full life cycle of such NoEs, which includes engagement, running (including renewal, redirection and support) and dissolution. Peter Sachsenmeier outlines that NoEs are relevant because they blend collaboration with competitive elements, and therefore become models for the truly valuable firm of the future. Concrete recommendations are given to make NoEs feasible; these include a hard look at the true benefits and risks associated with them. An overview of success factors for NoEs is provided. There is a need for accounting and NoE governance in order to establish an amount of management control over NoEs. The selection of suitable partners, shared goals, evaluation of the benefits for the partners and collaboration rules and regulations are of particular importance if NoEs are to succeed. NoEs will be sustainable if the environmental and work factors associated with them are assured, and teams properly supported. A substantial ewf programme management is suggested. »Network theory will take off if we keep our work multidisciplinary, comparative and international in nature.« The aim of NoEs is to »create value, wealth and new opportunities for everybody involved […] while reducing risks and costs.«

Discussion Forums

The discussion forums of the conference were the major channels through which, we had hoped, pertinent issues and declarations of interest in continuing work on the issues would emerge. The forums were structured so as to give the participants maximum opportunities for exchanges and communication. All forums reported back into the plenary through rapporteurs specifically provided for the purpose. The major issues were clustered and displayed to all participants, in the final plenary session.

In the event, we did not quite manage to achieve comparable results in all the forums. Other materials were distributed among the participants on the occasion of the conference. However, we present some contributions from the seven forums so as to provide insights into them and as a reminder of the ideas and topics raised in the discussions that took place.

The first forum, *Secure Engineering Collaboration*, attracted substantial interest; however, we have only a short summary by Martin Schmidt.

The second forum, *The Importance of Standards and eMarketplaces*, is represented by six contributions. Alexander Scholz introduces and summarizes the overall topic, Peter Mund writes on the discussion of electronic catalogues, Boris Otto outlines transaction standards, Jacques Spee deals with business process standards, and Roland Heidel together with Udo Döbrich attempt a generic reference model. Roland E. Wolff contributes a conclusion.

From the third forum, *Neutrality, Openness and Co-operation in the Service Industry*, Hagen Wenzek and Harry Teifel provide an elaborate analytical model, and results. Their forum conducted a controversial debate on the need of a neutral entity and other issues.

The fourth forum constituted an automotive block and ostensibly dealt with *Recognizing, Developing and Implementing Core Competences*. In essence, the forum continued to discuss the automotive presentations which had taken place in the morning, especially the idea of a Network of Automotive Excellence. This is amply reflected in the contribution by Reiner Speiser and Hans H Jung. The NoAE has since been become a major initiative and a corner stone of the work of ewf europe.

Forum five, with the provocative title *30% More Efficiency in Product Development*, is represented by three contributions. Rainer Kurek takes on the topic with the title of the forum, Reinhard Wagner writes on the *Productivity of Knowledge Work*, and Karl Dokter makes another contribution echoing the title of the forum.

Forum six had a substantial panel of presenters, and was of great interest to SMEs wishing to explore collaborative projects. Stefano Carosio gives an overview of *Virtual Level Collaboration*.

The seventh forum looked at Technology as Basis for Collaboration. Rainer Bugow, in *Technology as a Basis for Collaboration*, gives a succinct overview of the context, lists the problems and outlines the issues needing attention in engineering collaboration. Thorsten Reinheimer describes the *MOSAIC E-Net Integration Platform*, as one example of such technologies. Martin Schottenloher, in *Data Re-Presentation as a First Step Towards a Unified IT-Structure*, makes suggestions for a bridge between the gap separating EDM/PDM systems on the one hand, and ERP systems in enterprises, on the other hand.

Invited Contributions

The editors have invited a number of authors to make additional contributions which, we felt, were of significance to this book. These authors shed a - sometimes alienated and paradoxical - light on the issues, and we welcome their contribution to our thinking "outside the box".

Gunter Gebauer, in his *Competition and Collaboration in Sport*, shows that similar issues exist outside industry. Competitors become partners, and partners become competitors again. Manifestations of competition and collaboration change over time, and according to requirements.

Eric Scherer, a founding member of ewf europe, discusses the important issue of *Opening up to Change – People as Critical Success Factor for Collaboration*. His tenet is that action oriented designs of organisations which look at motivational issues right from the start lead to more sustainable success than approaches that target structure alone.

Stefan Kühl, interested in intrapreneurship, discusses *The Boundaries of Marketisation – The Myth about Entrepreneurial Employees*. He explodes some of our most cherished management myths, and shows the many paradoxical demands made on employees.

Anna-Maria Schäfer is a banker and writes on *Performance Community – New Financing Solutions*. She shows that, as OEMs and suppliers are increasingly connected, business models and each other's processes are deeply affected as well. A transfer of risks and problems takes place, including financial ones. Ultimately, this requires alternative and tailor made financing solutions.

Alexander Höppe deals with the *Identification of Cost Savings Potentials. Achieved by Collaboration using Lifecycle Costing*. Key to these are lifecycle management functionalities embedded in the support systems, and the identification of cost savings potentials through an effective product lifecycle costing exercise. One of the most important cost-saving potentials lies in collaboration with external partners, suppliers, and customers. Relevant tools can be used for information distribution, as well as for assigning and tracking tasks.

Klaus Kaplaner writes on *Competition and Collaboration in the Automotive Industry*. He notes that not only are markets saturated, but – at the same time – industry is confronted with the need for greater variety and more complex products. Collaboration becomes essential for survival. Ways to get there are outlined and discussed.

Oliver Merle of automotive association Odette describes an elaborate *Best-Practice Concept for Supply Chain Monitoring*. He provides a complex model, including rules for demand calculation throughout the network. He also addresses alert management, communications and workflow issues, multi-customer scenarios, and suggests key performance indicators as well as allocation of responsibilities for a supply chain monitoring system.

Martin Schottenloher, in *New Networks: Features and Tasks*, explores the issues related to the concept of new networks in particular in comparison with the structures in known alliances, networks and partnerships.

Ewf Information

In a short section devoted to ewf information, the ewf europe executive board consisting of Herbert Köpplinger, Franz Obinger, Peter Sachsenmeier, Sabine Schindler and Martin Schottenloher, describe their *Vision* for NoEs. Jürg Masson, president of the lively *ewf association in Switzerland*, an affiliated grouping, describes the philosophy and work of that regional group.

Finally, conference information including an overview of the authors, the conference programme and a list of participants round off this volume.

***

Changes as described in this book have a considerable psychological impact on the human mind. To the fearful they are threatening because they mean that things may get worse. To the hopeful they are encouraging because things may get better. To the confident they inspire because the challenge exists to make things better.

As editors and authors, we hope to have made you feel more confident in accepting the challenges of competition and collaboration in your professional life, in the networked industries of the future.

May 2003

Peter Sachsenmeier, Martin Schottenloher

President's Introduction: From Idea to Implementation

Herbert Köpplinger

President ewf europe

This book represents the official proceedings of the 2nd European Engineering User Conference (e.e.u.c. 2002).

The first e.e.u.c. took place in 2000 in the research and innovation centre (FIZ) at BMW, Munich, Germany; the third European event is scheduled to take place in November 2004 in Prague, the capital of the Czech Republic.

The Brussels conference focused on the theme "Challenges between Competition and Collaboration":

- How will companies manage the complexity between being in competition on the one hand and striving for partnership on the other hand in practical terms?
- How will companies position themselves strategically in new networks in order to survive?
- How do new forms of co-operation emerge – the so-called "Networks of Excellence (NoE)"?
- What tasks and roles re., employees and management can be adopted in these NoEs as a result of the enhanced complexity?
- What do new value systems in the network look like, what is the role of trust, openness and fairness?

The congress in Brussels was a new type of event for European top-management that dealt with the above-mentioned questions in detail. Next to seminal lectures from the automotive, the aerospace and the engineering and construction industries, current issues and problems were deepened in discussion forums and working groups. The two-day conference was rounded-off by contributions from the European Commission on future policies for manufacturing industry in an expanded Europe of ever more nations and increasing competition.

One of the main goals of the ewf europe is to formulate ideas and strategies for a Network of Excellence (NoE). Through the practice if NoEs, ewf europe hopes to shape and propel manufacturing industry forward to greater successes and who commonly develop advantage versus competitors.

To get there is not easy: Frederic Vester, the father of networked thinking, already noticed that the current education systems at universities and in companies focus on achieving easy, logical steps and on the definition of self-evident cause and effect principles. The current practice is still far removed from networked correlations in open systems and their non-causal behaviour. Unfortunately, practical solutions for NoEs are not yet available.

What role do trust, openness and fairness play in networks of excellence? Stephen R. Covey, the author of *The Seven Habits of Highly Effective People* and management consultant Reinhard K. Sprenger believe that trust creates more value than any other value enhancement concept can create. According to the philosopher Peter Sloterdijk the "society of the future" is condemned to trust. Trust turns into one of the key elements of modern economy and as such, is fundamental to NoEs. Many talk about trust, ostensibly it is the most important thing in co-operation – yet in effect it is the rarest thing.

And what does the balance of the Brussels conference, against the initial position, look like?

In the speeches and discussion forums, important and critical topics were discussed openly. Especially the representatives of the automotive and supplier industry appealed for fair co-operation among each other and essentially for partnership and mutual trust.

It was interesting to observe that, although individual industries such as automotive, aerospace or engineering and construction are completely differently structured at first sight, a number of common solutions and tasks appear. Therefore, the holistic ewf approach of the "Network of Excellence" for the overall manufacturing industry makes absolute sense.

After the Brussels conference, the foundation of a "Network of Automotive Excellence" is being discussed. Also, other contributions and discussion forums show initially promising approaches and results in topics such as "eBusiness and Standardisation".

We would like to thank all the organisers and everyone who contributed to the success of the conference especially our co-organiser AGORIA from Belgium, the European Commission, our marketing partners Hewlett Packard, T-Systems and Softlab as well as all the speakers and all those involved in the discussion forums as well as the participants for their enthusiastic contributions.

The congress has been very worthwhile. Now it is time to process the open issues together. In order to do so, these proceedings are meant as an initial contribution to make the issues and ideas discussed in Brussels available to a broader audience.

We would like to invite you to conduct a continuing discussion with us – across all industries i.e. the entire manufacturing industry, holistic (people – organisation – technology) and Europe-wide.

References

Vester, Frederic (2000) Die Kunst vernetzt zu denken: Ideen und Werkzeuge für einen neuen Umgang mit der Komplexität. Stuttgart, DVA

Covey, Stephen R (1989) The 7 Habits of Highly Effective People. London, Simon & Schuster

Sprenger, Reinhard K. (2002) Vertrauen führt. Worauf es im Unternehmen wirklich ankommt. Frankfurt, Campus

ewf europe (2002) Brochure 2nd e.e.u.c. 2002, München

Between Collaboration and Competition

The Future of the European Manufacturing Industry

Hans-Jörg Bullinger, Dietmar Fischer

Hans-Jörg Bullinger, President Fraunhofer Gesellschaft,
Dietmar Fischer, Director AGILeVIA GmbH

The future lies in collaboration across enterprises – it is essential in order to exist in competition

Increasing globalization, the growing pressure of competition which accompanies it and constant changes in the enterprise environment are placing new demands on the manufacturing industry. Worldwide partnerships, global networks and increasing outsourcing of development output influence and modify the market on an enduring basis and force enterprises to react ever more quickly to the resulting demands. More than ever, process and product quality as well as an effective and efficient organization of the process follow-up along the entire life cycle of the product determine the continuance, growth and success of enterprises, product branches and, in the end over the future of entire fields.

Develop new markets, promote innovation skills, shorten time-to-market and lower costs- challenges which will in the future be solved between "collaboration and competition", in strategic partnerships and collaborative networks (Fig.1.: Accepting the challenge).

The future of the European manufacturing industry lies in collaboration across enterprises over the entire added value chain. The prerequisite for working added value chains is the potential of the enterprise for continuous digital support and digital representation of co-operative enterprise processes as well as complete command of the complexities of product and process. Those who can supply the digital factory will survive.

Fig.1: Accepting the challenge

New orientation of the automobile industry: challenges between collaboration and competition

The driving force of global added value is the automobile industry. It needs to find innovative forms of collaboration and has recognized the value of strategic partnerships and collaborative networks from an early stage. The collaboration of OEMs, suppliers and people employed in development is becoming the key factor for joint competition ability (Fig.2: Securing success in competition).

In co-operations and enterprise associations there is an attempt to constantly improve the communication of supplier and customer and to bind external engineering partners into the product development as early as possible. Numerous examples of strategic alliance collaboration across branches, joint ventures but also the co-operations of competing enterprises speak for the success and necessity of collaboration across enterprises.

Co-operative, virtual, integrated product creation becomes the enterprise strategy, a development which aims for the creation of flexible enterprise networks and added value chains across organizations and branches.

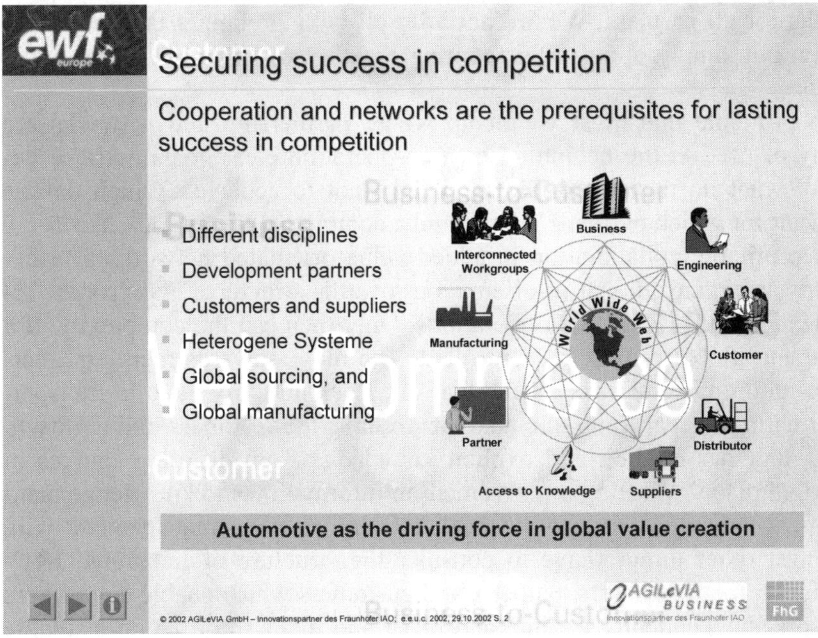

Fig.2: Securing success in competition

In order for a collaborative network to function, flexible organization structures are needed which can, circumstances permitting bring together the widely scattered fragments of the network. An IT infrastructure which ensures that all relevant informations flow securely and smoothly in real time between all involved is needed- without a doubt technologically possible. The transformation into a collaborative business network, a concatenation across the enterprise limits, is a difficult, but not impossible challenge for enterprises. The required integration efforts must however happen step by step: first internally and then externally. In the first step of the realization of the concept of collaboration, enterprise processes must be imaged internally and optimized. Systems and operational applications which do not yet co-exist must be placed on a uniform platform.

An infrastructure aimed at cooperation which enables the integration of external partners into projects by means of modern I&C technologies can however only the first step. The qualitative information and knowledge exchange inside individual enterprises but also between different business partners must be intensified further. One needs to develop innovative business models which take into account the enterprise strategies of all involved- after all, co-operation can only work successfully if everyone involved profits from a networking and a future securing win-win situation results for all partners. We are not after all talking about a take-over or a fusion but simply about a fair coming together of enterprises with equal rights.

In principle one must weigh up where partnership and networks are really of use. At the beginning of a co-operation clear goals must be defined: what do the enterprises involved want to achieve? Which data is relevant for which partner? What could a contractual layout look like?

A profitable collaboration in added value orientated networks also demands an organizational re-orientation towards structures and process organizations beneficial to co-operations. Only then can the complexity of a large and global project be controlled. The move towards merging enterprises and entire markets means that one does not only have to transport information in real time, one also has to have the readiness and ability to bring together implicit and explicit knowledge from different sources in the right place. A lasting improvement in information and knowledge management must not fail to be realized. Information management will, amongst other things, have to consider the structure of integrated I&C-platforms for co-workers, partners and customers which enable co-workers to work independently of time and place and make information available according to the team and the project. Enterprises which want to stand their ground and differentiate themselves in international comparison will have to improve their ability to expand knowledge further and to develop knowledge and furthermore to use their knowledge systematically for innovation instead of only, as has been done till now, conserving it.

Original equipment manufacturer: concentration on core competences

The general intensification of competition in the medium-term leads to the concentration of OEMs. In the case of rising demands improved performance, shorter development times and sinking costs are expected of enterprise areas. A development, which leads to automobile manufacturers concentrating even more intensely on core competences – development, production, integration and assembly, the classic core competences of automobile manufacturers are becoming increasingly unattractive as they cause high costs. Apart from the focusing of core competences, new core competences must at the same time be defined and built up. In order to open up new channels of added value it will be essential to integrate market and customer management (mass customization and order to delivery concepts) strongly and to consequently expand the range of services on offer (see Fig.3.: Secure competition success, but…).

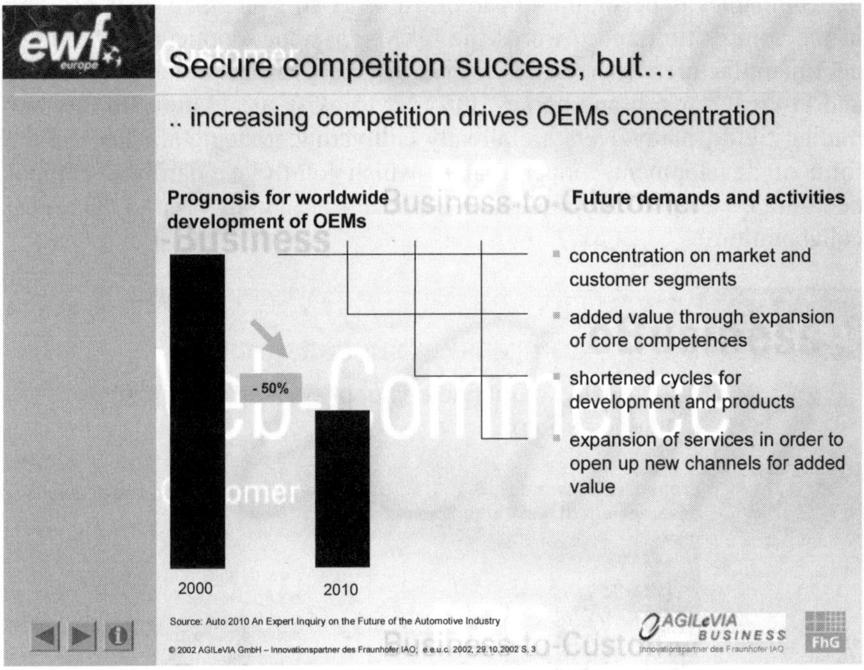

Fig.3: Secure competition success, but…

This is an orientation which demands flexible organization structures in which the enterprises can proceed effectively and efficiently. The developing change of product driven enterprise strategy (push-strategy) to customer need and service orientation (pull-strategy) in no- production areas underlines the tendency of branding. Products will in future no longer be marketed over quality characteristics such as “Made in Germany” but independently from the production site over attributes such as “Made by DaimlerChrysler”. A successful model of trend-setting production and enterprise strategy is the Smart production in the network. The small car with a manufacturing depth of less than 10% is composed of components which are for the most part pre-manufactured by the suppliers integrated in the production line. The grouping of product output and service is completed by the Smart complete offer (financing, leasing and insurance offers).

Supply industry: rising demands under increased competition

The rising demands under increased competition will also have the consequence of a market clean-up in the supply industry. Take-overs and amalgamations will increase further. A trend towards the development of Tier 0.5 companies is becoming apparent. In order to strengthen their position in the competition and towards the OEMs, system suppliers and component manufacturers increasingly strive to take over technology leadership and to create own brand names. In order to exist successfully in the new trading fields, many Tiers are already cultivating strategic alliances in the form of development cooperation in which jointly standardized components are developed and production capacities are jointly used to maximize collaboration.

Fig.4: Secure competition success, but ….

Smaller and smaller quantities with a larger number of variants and shorter life cycles- thus, shorter reaction times and highest flexibility in production in the future become, more than ever, guarantees of the ability to compete. Logistics is becoming the deciding factor in international competition. The development of the usual logistics chains to quick and flexible logistics networks must be lastingly supported. The realization of effective collaborative supply chain management is needed in order open up performance potentials across enterprises with the concatenation of added value orientated processes. In order to achieve far-reaching improvements, collaborative supply chain management must already be used in the planning phase so that all involved enterprises are pulling at the same string from the beginning (Fig.4.: Secure competition success, but…).

Collaborative engineering in added value orientated networks

The success in competition in the technical product market is dependent of time, costs, quality and flexibility. The market demands innovative products coordinated to customer wishes in a large number of variants. The prerequisite for the development of innovative products is an efficient product development. In order to already operate in a modern and cost-conforming way in the early product creation process, where the greatest amount of the product costs is still decided, the product development must be parallelized and integrated organizationally as well as from the point of view of the information technology. A collaborative, virtual, integrated product creation must be declared as the strategy (Fig.5: Challenges for engineering).

The good experiences and results of the Siemens AG speak for the success of collaborative engineering models. For a long time the enterprise has been profitably practicing co-operative "follow- the-sun" engineering around the world, around the clock. The distribution of the development structures in different locations supports the concentration on core businesses und in addition makes it possible to take advantage of location costs.

Defined processes between all the involved and a global information and data stream are decisive for an effective co-operation in engineering.

In order to ensure full communication and integration of all the products and processes it is essential to integrate all partners into the product creation process and synchronize production networks. Communication and information streams must be optimized to such a point that a quick and tight inclusion of external engineering partners is already ensured in the early development phases.

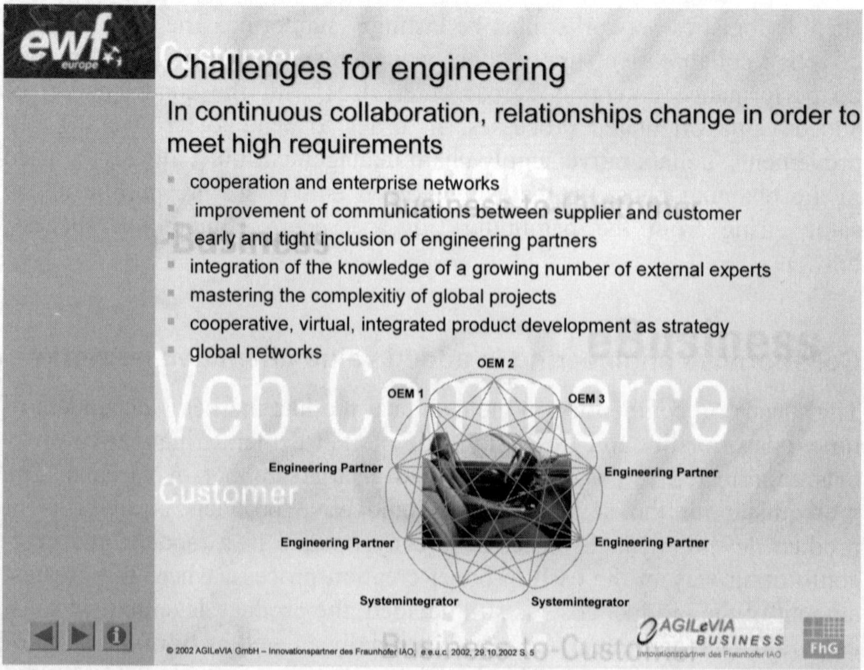

Fig.5: Challenges for engineering

The integration of the knowledge of the increasing number of external experts is an absolute necessity. The bringing together of distributed know-how increases the reaction speed of organizations and contributes to the efficient grouping of resources. The prerequisite of mastering global and complex projects is the implementation of a collaborative project management which supports a project-orientated technology development and makes a high transparency of decision processes possible. The increasing complexity of products is threatening to drive up the development times and thus the project costs severely. Therefore, the demands on the project management rise considerably. Extensive and strict quality standards are becoming essential. By means of collaborative, web based applications, a transparent project management can be standardized in which all involved in the process constantly have actual, transparent and measurable information at their disposal. The constant checking of goal achievement and the quality status in development projects allows for early, flexible intervention in the construction phase and reduces cost-intensive changes in later processes, a lesson which DaimlerChrysler had to painfully learn at the introduction of the A-class (elk-test). In order to improve internal communication and the co-ordination of development, production and marketing during the development time in future, and to recognize possible during the development time in future, and to recognize possible signs of product weakness and repair it, the group successfully introduced so-called "quality gates". Quality gates support a transparent product development including of all involved in the project: individual milestones of a project are summarized along the added value chain in one quality gate. In order to pass through a quality gate, defined quality results must be achieved at individual development and marketability levels. The collaborative, transparent product creation process means unambiguity and liability for all involved in the process. The result, next to a quicker and more efficient approach, is the marked increase in the degree of product and process maturity while avoiding unnecessary iteration loops.

Synopsis

The opportunities of the globalization of the competition lie in markets which are to be opened up, logistics systems which operate worldwide, consequent information networking, homogenized products and new forms of job sharing- potentials which must be uncompromisingly opened up and supported across the added value chain (Fig.6: The future of the manufacturing industry). Those who act with suppliers, customers and partners in global competence and co-operation networks and decisively use the competition advantages of innovative technologies and products will be successful.

Fig.6: The future of the manufacturing industry

ERA: A Challenge for Co-opetition

"Towards Knowledge-based Enterprises"

Hervé Péro

Head of Unit – DG Research – European Commission

The European Councils of Lisbon and Göteborg have clearly set the objectives of research activities: to help the Union to become the most dynamic and most competitive knowledge-based economy within 10 years, and to support sustainable development (environment, economy, employment) in the context of social cohesion.

The industrial point of view… a need to evolve

European industry is faced today with many challenges to remain competitive in an increasingly complex environment, characterised mainly by the word "globalisation". The speed of innovation, pushed by the consolidation of the Information and Communication Society, and the rapid uptake of new technologies, such as nanotechnology, is counterbalanced by the growing cost of high level research. The societal role of industry is also being recognised, in particular regarding sustainable development and the ethical and social consequences of economic and technological evolution, notably considering human resources and employment prospects.

These challenges force European industries to evolve, modernising their manufacturing approaches, i.e. working much more in networks, and adapting their structure, strengthening links between research and innovation, increasing added value of products, production and services, and of course decreasing their environmental impacts. This can be characterised by a transition from resource-based to knowledge-based organisations.

For this evolution to occur, research is becoming a key and unavoidable issue.

Many industrial infrastructures, production sites, facilities and distribution networks will have to be reorganised. The twofold transition towards knowledge-based society and sustainable development demands new paradigms of production and new concepts of product-services.

Research is crucial to help companies modernise and improve competitiveness. There is a need to move from resource-based approaches towards more knowledge-based ones, from quantity to quality, and from mass produced to higher added value and eco-efficient products and services.

In this respect, integrated research activities become key factors to support transformation of the European industry towards more responsible organisations. This highlights also the importance of new approaches, such those fostered by nanotechnology, biotechnology, information technology, knowledge-based materials, and many other new processes, as well as the effective assessment of their impacts over their whole life-cycle.

Research is needed … but should shift existing paths towards sustainable development

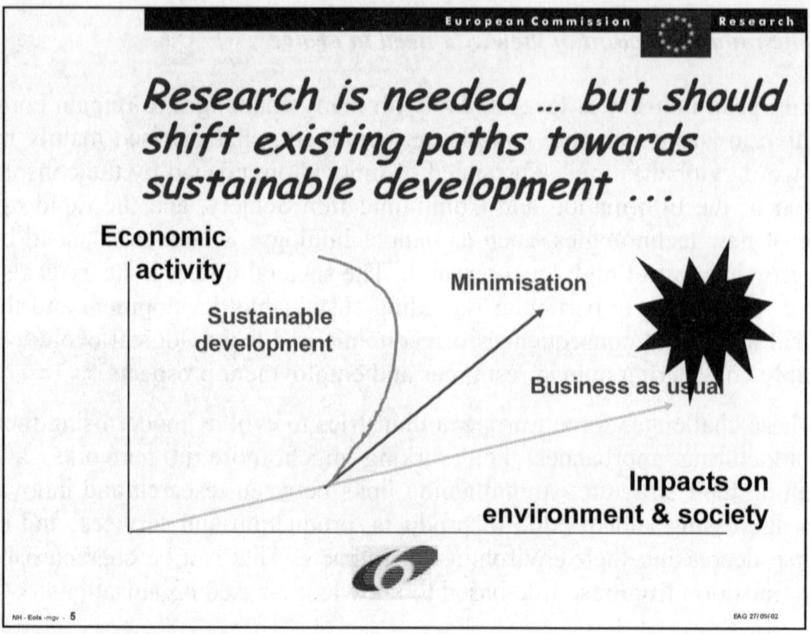

Fig.1: Sustainable development

Indeed the "business as usual" approach and even the efforts towards minimisation of impacts on environment are dramatically leading in medium to long term to a huge crisis at world level. The only solution is to turn towards sustainable development, which means ensuring the same - if not better - conditions to the next generations as we have today, without negative impact on our economic activity. Sustainable development may be a utopia… but what a stimulating goal for researchers and industrialists!

The 6th research and innovation Framework Programme

The 6th research and innovation Framework Programme of the European Community is characterised by a changing role of EU Research Programmes. It is based on three clear "political" pillars:

- The "Lisbon" objective has been stated in spring 2000: "to help Europe becoming the most dynamic and most competitive knowledge-based economy within 10 years".
- The Göteborg" objective, stated in spring 2001, puts sustainable development, i.e. environment, health, economy, employment, at the top of the agenda of all EU activities.
- Finally, the development of a true European Research Area (ERA) is the driving force of the EU research policy. Integrating, reinforcing and structuring research efforts at European level will be the objective of the 6th Framework Programme. In addition, such an effective area should stimulate industrial investments in research, thus closing the loop with the "Lisbon" objective. It should be reminded that Europe strongly lags behind its competitors in terms of investment in research. The objective is to reach 3% of investments in research in Europe at the horizon 2010.

The 6th Framework Programme is concentrating its efforts towards 8 research priorities, which support competitive and sustainable growth (see Fig.2).

The future of manufacturing industry is directly or indirectly linked with the eight priorities of FP6. Sustainable competitiveness of industry is for example clearly linked with the deployment of information society technologies, but in a context of increased human- and environment- friendly approaches. Future products will more and more be associated with eco-efficient services, emphasising the importance of sustainability. Health and security related issues are also becoming of prime importance and will generate growing markets in the coming years, influencing in consequence the manufacturing industry. All these consideration will affect the sector of SMEs, which represent more that 99% of enterprises in Europe. Special attention will be devoted to them in FP6.

Fig.2: The eight research priorities

Fig.3 below shows how the priorities have evolved during the last 20 years. The emphasis is clearly more and more towards societal issues.

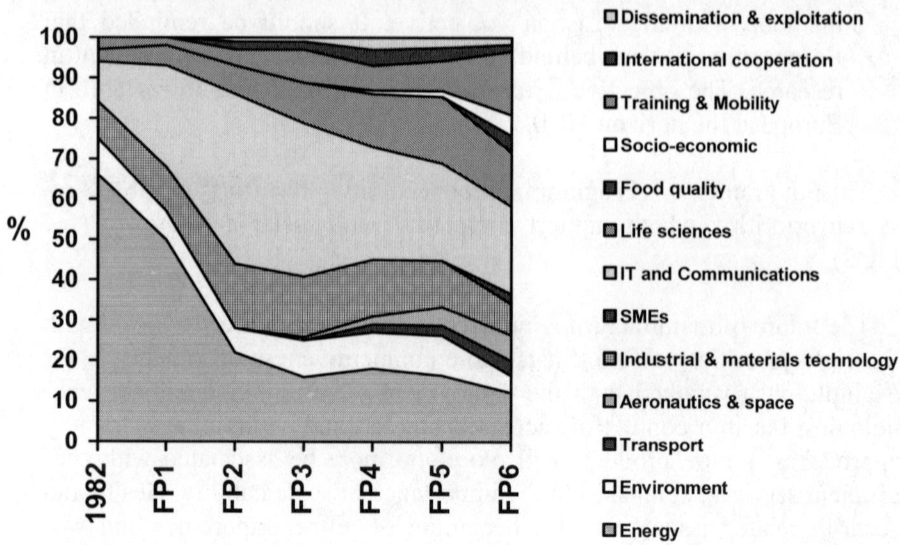

Fig.3: Evolvement of priorities during the last 20 years

As far as industrial technologies are concerned, their importance has steadily grown over the years, with some specialisation recently (e.g. referring to the "transport" sector or to SMEs).

The enlargement of the Union... an opportunity not to be missed

As far as the international dimension is concerned, it is clear that enlargement of the European Union will be a key challenge, not to be missed.

The 6th FP will be the first action from the EC where the accession countries will be considered at the same level as the Member States.

A lot of potential exists for the manufacturing industry, not only because of a new market, but also mainly because of the new technological opportunities. Indeed the culture of these countries will bring complementary views to those developed up to now by EU industry. Not to forget also the need to modernise the industry of these countries...

The importance of SMEs in Europe

Support to SME aspects is important. There is a strong need to develop further the infrastructures that implement new technologies and organisational practices. Europe wide networks are required that give the smaller companies also access to new possibilities in product and process technologies, therefore stimulating implementation of paradigm shifts in traditional sectors dominated by SMEs.

The manufacturing industry in Europe covers various sectors, mainly concentrated in more "traditional sectors". CRAFT and Collective research will therefore be important tools that will be implemented under FP6.

Apart from these actions, at least 15% of the total budget allocated to the priorities of the framework Programme will be dedicated to SMEs. Large possibilities will exist to include high-tech SMEs in consortia, or to consider specific "modules" for SMEs in future projects (e.g. 15% of budget) and / or to launch specific actions and measures.

In this context, the role of industrial associations, e.g. EWF, should be highlighted to diffuse information, stimulate participation, and ensure dissemination and good use of results.

There is a need to develop efficient knowledge management in Europe

The challenge is to support the development of a high-tech industry, based on an effective, knowledge-based workforce. Knowledge management will be a key enabler of competitive advantage!

As mentioned the Lisbon strategy implies a reinforcement of knowledge management for efficient use by EU industry. However, the facts show that in the '90s, EU was leading in scientific output (cf. amount of publications) but not in technological output (cf. amount of patents).

This highlights the need to strengthen further the links between research and innovation, those between fundamental and applied research, and the synergies to be gained in developing efficient relations between industry, research and education.

Immense efforts are needed by the EU industry to invest in research to be able to bridge the potential knowledge gap with their competitors

What is particularly crucial is that European industrial research lacks human resources ... The following graph shows that EU industry employs less than half the number of researchers than the competitor's regions:

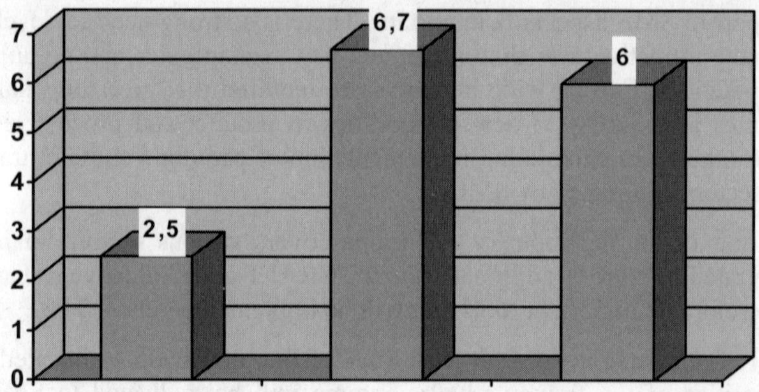

Fig.4: Number of researchers in firms per 1000 labour force (1997; EU, USA, Japan)

The new instruments of FP6 will help ...

Recent Councils have confirmed the importance of research and innovation policies and of developing a true European research community, through integration, structuring and reinforcement of related activities.

This means wider scope and ambition than in the past. This also means that the new instruments, the integrated projects and networks of excellence, should support more effective research activities.

The new instruments in FP6, namely Integrated Projects and Networks of Excellence, will allow better co-ordination and effectiveness of research activities at EU level. Several objectives should be encompassed such as ensuring system competitiveness and sustainability, quality of life, sustainable employment, high quality education, and of course ethical aspects, in an integrated approach at European (even international) level. Research and innovation activities should be jointly present in projects to ensure the quick exploitation of research results. Education and skill development should receive particular attention in view of the challenge of the "knowledge-based society" and "knowledge-based industries".

Integrated Projects will focus on clear and quantifiable objectives.

Integration issues should be tackled at least on four aspects:

- multidisciplinarity of research
- integration of research, innovation, awareness and skill development activities
- integration of partnership along the value chain and/or for creation of knowledge communities
- complementary funds from different sources, private, regional, national, European

This involves ambitious, far-seeing research actions and effective research projects to rapidly spread results into practice. It involves also a strong presence and interaction of innovative enterprises and research organisations, from all Europe, in research consortia.

Such a challenge implies the acceptance of large and complex networks. It implies therefore that companies should accept sharing of knowledge and *co-opetition*, i.e. co-operation in research and innovation activities, while continuing competing on the market place.

Networks of Excellence are another response to this need for networking. They will be promoted to assemble EU research capacities and strengthen S & T excellence for the competitiveness of EU industry. Partnership of NoEs will be characterised by participants wishing a progressive and lasting integration of their research capacities.

Final remarks

Very often, engineers think about engineering, organisation, and technologies… and tend to forget the overall long-term objective of industry, which is to create wealth and provide well-being for the society:

- The knowledge society of tomorrow implies the transformation of industry towards knowledge-based organisations.
- In addition sustainable industrial production, is the challenge to keep in mind! It involves re-design of existing products and processes, as well as a "team work" at European level.

In view of the efforts needed to support these transformations, it is clear that nothing will be possible if research and innovation remain limited at an individual scale. A true *European Research Area* is crucial to help these transformations to be as effective as possible.

In conclusion, and as far as concerns the directorate general for research of the European Commission, the main messages to ensure the future of manufacturing in Europe are the following:

- **Increased efforts towards research:**

  The EU industrial sectors need it!

- **Integrated efforts towards strategic goals:**

  Research without innovation is of no value

- **Co-ordinated efforts in research:**

  To ensure synergies between EU and national actions

- **Concentrated efforts:**

  Towards sustainable development and knowledge-based enterprises

For more information about the Community research and innovation framework programme, please consult:

http://www.europa.eu.int/comm/research/industrial_technologies/

Plant Engineering in Global Competition

Ludolf Plass

Chief Technology Officer, Lurgi AG

The plant capacity of what today is referred to as "world scale" for almost all products of the organic and inorganic chemistry is two or even three times higher than what was considered mega capacity still 6 years ago. In parallel, the time leads available for modifications and new designs are shrinking.

An increasingly decisive factor to remain competitive in this field will be to provide turnkey projects including the necessary financing as a general contractor. The demands on the size of plants, their efficiency and complexity is bound to increase in order to consistently counter the cost pressure of global markets with constantly improved plants.

"The latest state of the art in technology at the lowest price within the shortest period of time turnkey at a fixed price". *This is the Global Challenge for plant engineering contractors at the outset of the new millennium.*

Mastering this challenge requires Global presence for market and sales reasons and "system leader" capability for the world wide execution of large scale contracts including a complex network of subsidiaries, consortium and / or joint venture partners, sub-suppliers and manufacturing sites, all based on uniform quality standards.

A recent benchmark has shown an average of 3 - 6 % error cost in plant construction. Quality improvement programs have been initiated by many contractors to substantially reduce error costs and hence quality costs *on the basis of selective re-engineering and training measures.*

Consequently, substantial changes in the workflow *and contents of the business process of the value claim in plant engineering were necessary. A consistently workflow orientated engineering process, based on engineering documents, standards and work packages and supported by uniform DP-Systems, mainly 3 - D PDS has been installed.*

The engineering and construction of a very large Methanol plant incorporating various Engineering partners and many sub-suppliers demon-

strates as an example the efficiency and complexity of state of the art plant engineering.

In addition, value added engineering *must contribute to continuously improve the technical and economic advantages of the overall plant.*

Value engineering is used o reduce investment and operating cost. Process intensification yields higher efficiency of process steps. R+D in combination with process simulation reduces development time and thus time to market.

Micro reaction technology will play an important part in the future development of plant engineering.

Typical examples illustrate the potential of these methods.

A comparison between the product development process of BMW and the plant engineering process of Lurgi shows more similarities than differences. Key aspects will be discussed.

Introduction

The Title of this conference is: *Challenges between Competition and Collaboration* and the subtitle is: *The Future of the European Manufacturing Industry.*

As the subject of my presentation is basically as contrary as possible to the subtitle of this conference, I would like to start with the following introductory remarks.

1. Keyword 'Manufacturing Industry'
   Lurgi is an engineering company without any manufacturing facilities of its own
2. 'European Manufacturing Industry'
   90% of Lurgi's sales are generated abroad, 70% outside of Europe.
3. Product Portfolio:
   Lurgi engineers and builds plants for the petrochemical and chemical industries as well as for the segments specialty chemicals, pharmaceuticals and renewable resources. All our plants are tailor-made. In contrast, one of the main products of the industries represented here is the automobile, a typical serial product.
4. Value of the Products
   An automobile typically costs € 50,000, a plant € 100 to 500 million.
5. Conclusion:
   Why does it make sense to give a presentation on the subject "Plant Engineering in Global Competition" despite these apparently substan-

tial differences, even oppositions? First, let me say the following: In our group, we have held discussions on subjects like efficiency improvement in plant engineering and shortening of product development times on several occasions with our top management. In this context, reference was often made to the automotive industry's success in process management and in the shortening of product development times. We, too, were convinced that there could be nothing in common between both industries. A benchmark with BMW on the shortening of development times in 1999 showed, however, that there are of course considerable differences but also substantially more common points that we previously believed. I will refer to this in more detail at the end of my presentation.

Market Development

According the German VDMA 2001Status report, the volume of new orders in the plant engineering and contracting business has substantially increased in 2001 in comparison to the year 2000. This rise has been especially strong in the Near East, the Asia-Pacific region and North America. (Fig.1) Also, in Eastern Europe /CIS, sales have increased but still remain

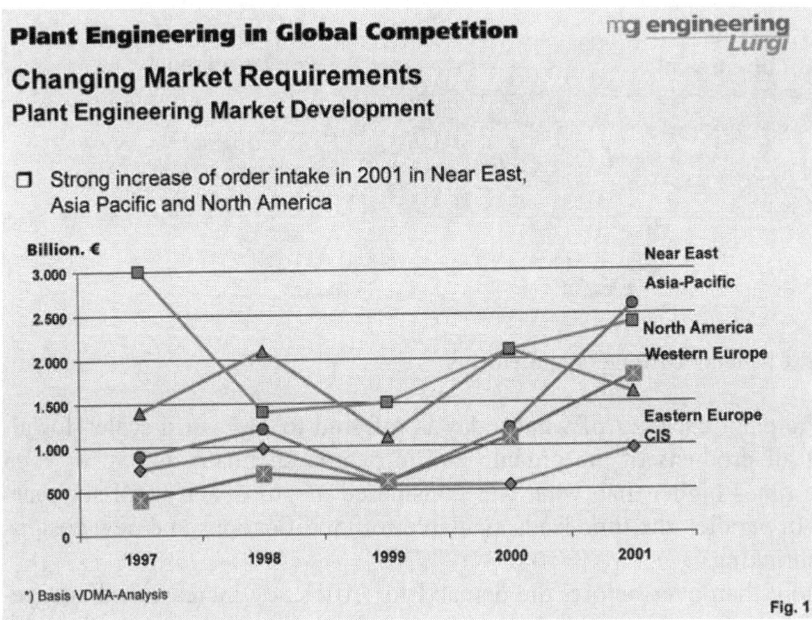

Fig.1: Changing market requirements

on a low level, whilst the industrial Western European countries show a considerable reduction in new orders.

There is no doubt, however, that customer demands have changed substantially. An increasingly decisive success factor to remain competitive in this field will be the ability to provide turnkey projects including the necessary financing as a general contractor including, in specific cases, equity participation and the operation and service of plants (Fig.2). Demands on the size of plants, their efficiency and complexity are bound to increase in order to consistently counter the cost pressure of global markets with constantly improved plants.

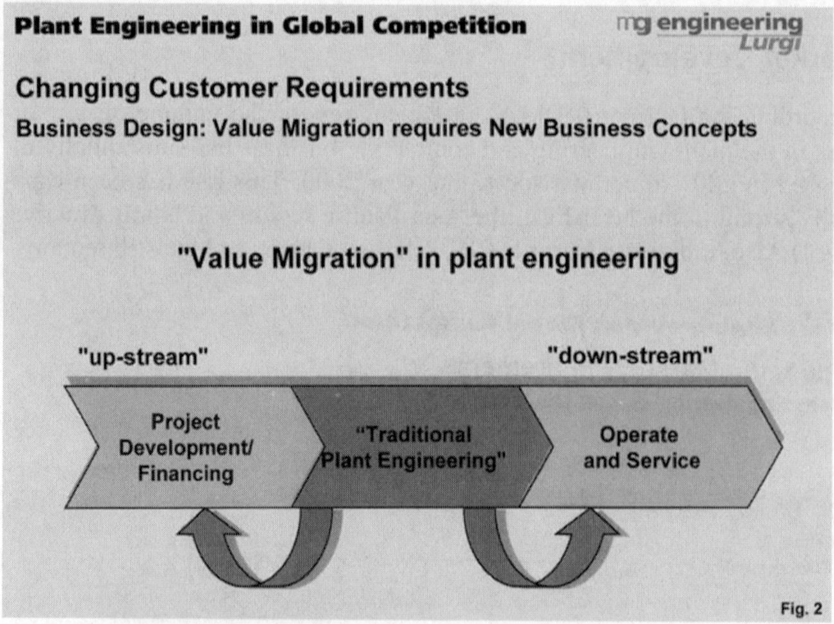

Fig.2: Changing customer requirements

The plant capacity of what today is referred to as "world-scale" for almost all products of the organic and inorganic chemistry is two or even three times higher than what was considered mega-capacity still six years ago. In parallel, the time leads available for modifications and new designs are shrinking.

More than ever before, the demand for efficiency increase and cost reduction presupposes technological leadership because it is only with an integral, technology-based approach that the necessary improvement in prod-

product quality can be achieved while at the same time cutting product life cycle costs.

"The latest state of the art in technology at the lowest price within the shortest period of time, turnkey at a fixed price". These are the global challenges for plant engineering contractors at the outset of the new millennium.

Document Orientated Workflow

Essential preconditions for mastering this challenge are, primarily, global presence for marketing and sales purposes and "system leader" capability to acquire contracts worldwide with the involvement of subsidiaries, consortium and joint venture partners, based on uniform quality standards. Whilst in the past the basic engineering was essentially done at one place and the detail engineering was commissioned to third parties in the sense of an "extended work bench", globalization of the business requirements and shorter cycle times now call for a substantial change in the workflow and contents of the business process of the value chain in plant engineering (Fig.3).

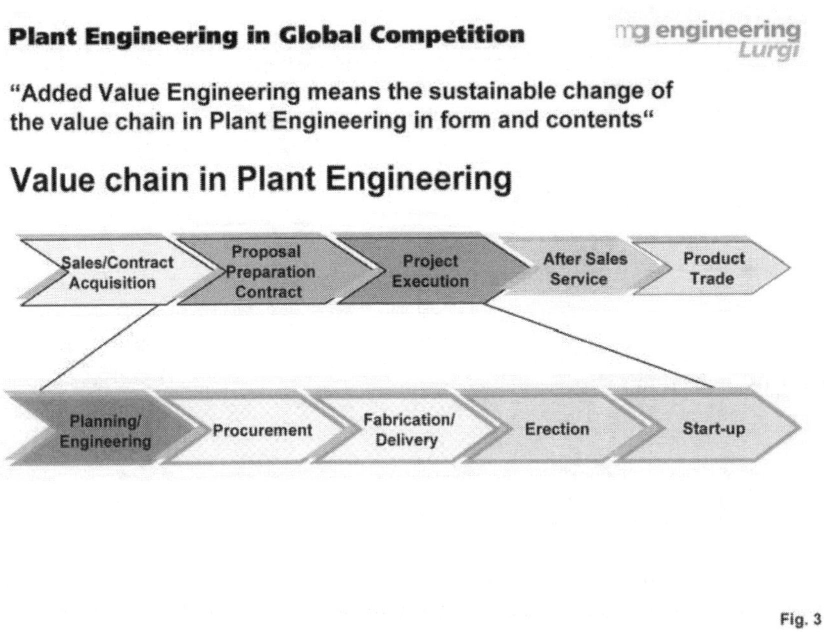

Fig.3: Value chain

So far, different, mainly individual software systems have been combined, mostly by manual data transfer. In spite of all quality management systems in place, in practice quality assurance was invariably based on an informal network of the implicitly available know-how of long-standing experts. Almost everything was engineered in-house, components were bought out - in Germany for the most part.

Optimization of the engineering business process requires a workflow standardization on the basis of the engineering documents to be prepared, a clear definition of modules, standards and work packages, and consistently DP-based job execution structures for a workflow-oriented, uniform engineering process. Fig.4 shows the fundamental structure of this document-related workflow.

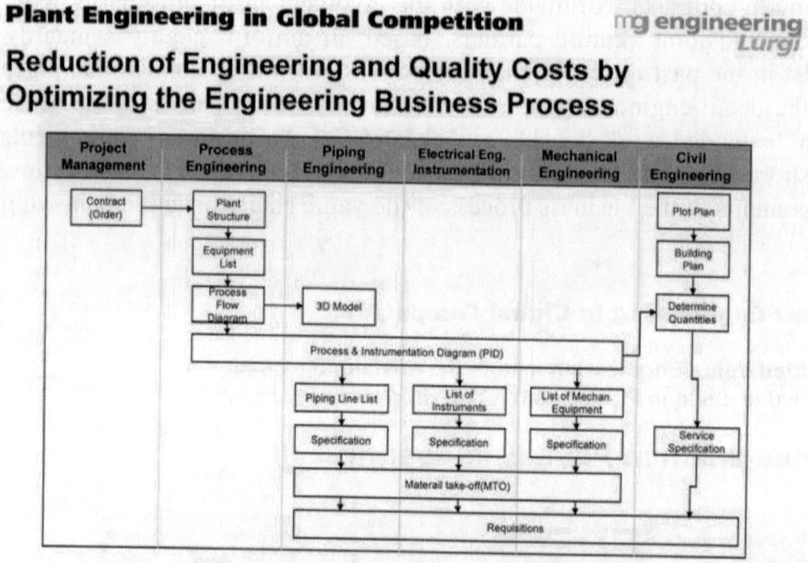

Fig. 4

Fig.4: Cost reduction

Worldwide Engineering

This change in the engineering approach shall be demonstrated by the methanol plant Lurgi has recently built in Argentina. A 3-D PDS system was used for worldwide job execution. The basic engineering was done in Frankfurt with the involvement of key personnel from Lurgi Bipronaft in

Poland. The dimensioning and basic allocation of the essential process equipment units were developed in the 3-D PDS system (shown in (light) *gray* color, in Fig.5).

Fig.5: Reducing engineering costs through global engineering

Fully engineered subsystems from a Brazilian strategic sub-supplier (shown in *black*) were also incorporated. The further detail engineering work was done under the management of Lurgi Bipronaft with local support by key personnel from Frankfurt. The structural steel engineering was tied into the PDS system by a supplier from the Black Forest (*dark grey*), all foundation and construction activities were contributed by the Argentine consortium partner, Techint. The complete piping engineering (*white*) was carried out at Bipronaft. Weekly, sometimes even daily calculated clash reviews and interface checks made sure that this form of "concurrent engineering" led to quality-conforming, integral engineering.

So the necessary prerequisites for successful global engineering are as follows:

- a clear job execution strategy

- a strong project management
- well-trained personnel
- control of interfaces
- uniform tools
- DP-based simulation tools
- worldwide purchasing to enhance global sourcing, increasingly based on e-commerce solutions
- worldwide information ≙ knowledge management
- controlling, risk management (including early warning) and quality assurance systems geared to global plant construction

Quality Cost

A benchmarking survey performed by management consultant Arthur D. Little (ADL) in 1998 with respect to error costs in plant construction (see Fig.6) shows an average of 3 – 6 % error costs. This does not include the costs for the generation of quality, which means project-independent and project-specific testing and inspection costs.

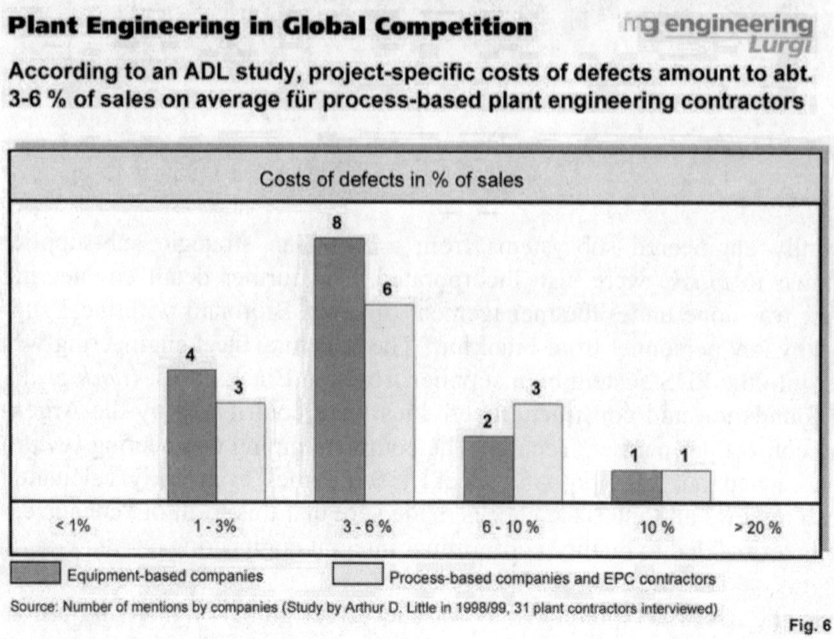

Fig.6: Project-specific costs of defects

Consequently, the error costs are in the same range as profit margins targeted by renowned plant contractors, but which in reality are hardly achieved worldwide at present. A benchmark study of the German plant contractors Linde AG, Krupp-Uhde AG and Lurgi Oel·Gas·Chemie GmbH led to comparable orders of magnitude regarding error costs. As a consequence, all three companies have set the task to substantially reduce error costs and hence quality costs on the basis of selective re-engineering and training measures.

Value Added Engineering

In order to be successful for the benefit of our customers in global plant construction and thus to achieve and maintain the reference position and customer loyalty which are decisive factors for this industry, plant contractors have to act as partners of their customers by increasingly making own material contributions to economics.

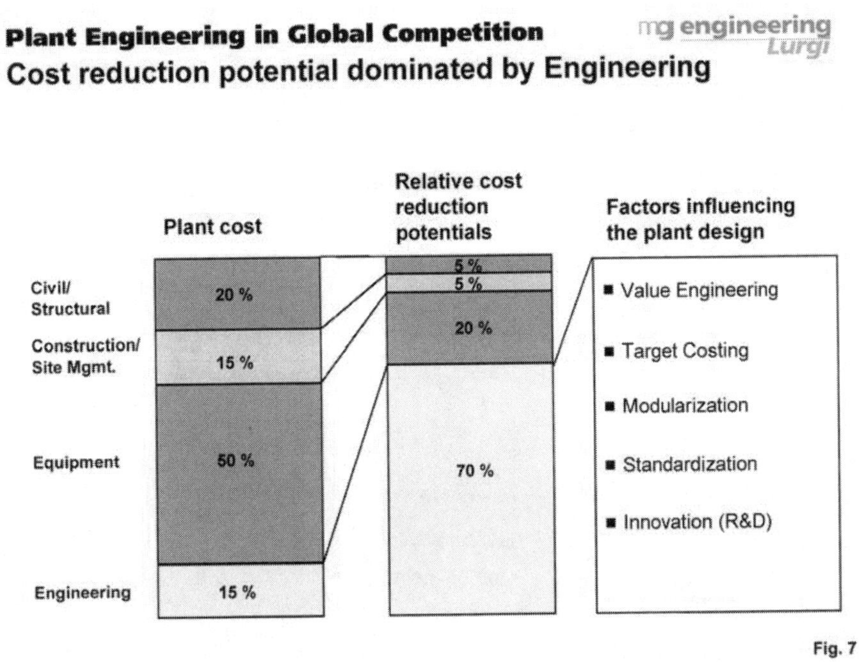

Fig.7: Cost reduction potential dominated by engineering

This includes proprietary processes and their optimization as well as new developments through ongoing innovation efforts in the same way as value engineering to reduce the costs of existing processes, efficient leverage of methods for target costing, modularization and standardization.

An analysis of the savings possibilities in plant construction shows (see Fig.7) that the cost curbing potentials are dominated by the engineering part. Although engineering only accounts for 15 % of the total plant costs, it defines in fact around 70 % of the cost reduction potentials.

Value engineering means in this context reducing functionality and equipment to the essential minimum required to obtain an operative plant. This can be achieved by rendering the process steps and equipment items leaner, but may also include alternative concepts in terms of process and equipment. Practical examples of plants implemented have demonstrated that savings of up to 30 % of the capital investment are attainable in this way.

A further essential element in Global Engineering is the so-called process intensification. This is defined as the cost reduction of products from plants in which distinctly more favorable production costs are achieved as a result of optimized chemical/physical process control in innovative reactor systems.

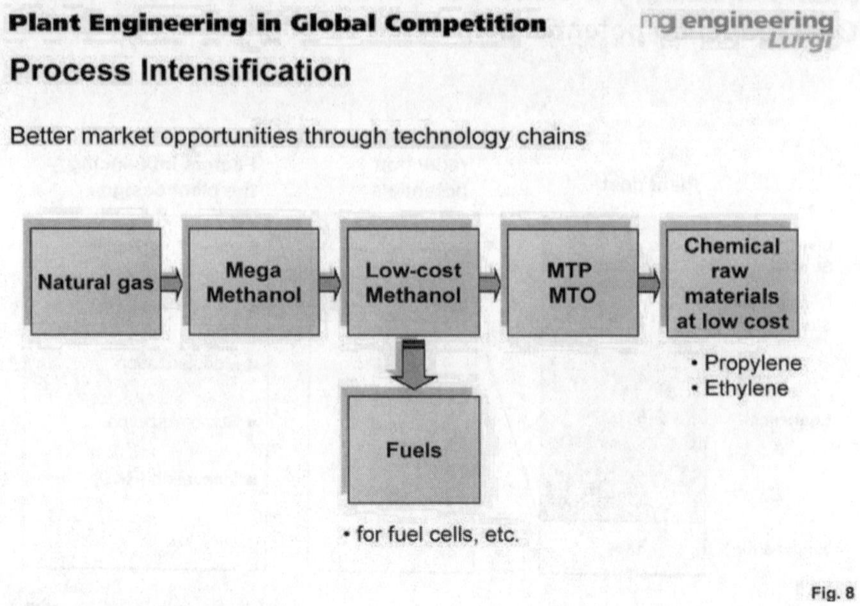

Fig.8: Process intensification

A typical example from the industrial plant construction sector is the production of methanol from natural gas according to the Lurgi Mega-Methanol process. Single-train doubling of plant capacity from formerly

2,500 tons per day to 5,000 tons per day and above is accomplished through process intensification. The use of oxygen instead of steam and an innovative reactor concept are lowering the methanol production costs by about 30 %. This opens up extended applications of methanol for the production of high-grade petrochemical intermediates such as ethylene and propylene in the same way as its use in fuel cells (Fig.8).

Micro-Process-Technology

In conclusion there are some critical questions: does the globalization route of world-scale plants, exclusively, or almost necessarily, lead to ever larger plants? I don't think so: if the total cost of a product for the end consumer is analyzed, it can be seen that the handling, storage and selling costs from large plants are becoming so significant in relation to the cost of production in the large plant itself that locally installed smaller plants are getting more economical for the customers if the raw materials price is favorable. This may apply in particular if the potentials inherent in value engineering and process intensification are exploited consistently.

Fig.9: Oil rig

Micro-reaction technology is one such consistent application of the methods of process intensification. Whether in the foreseeable future micro-reactors can be used to produce chemical commodities on the principle of "numbering-up instead of scale-up" (see Fig.9) is a question that will certainly still require intensive research and development efforts, but there is no doubt that important potentials of such methods like

- high-throughput screening (e.g. catalyst development)
- efficient tools for process development
- in-situ production / processing of specialty chemicals / pharmaceuticals

will find their way to be applied in chemical engineering and, consequently, also in plant engineering.

Benchmark between BMW and Lurgi

The required continuous and rapid improvements in plant engineering can only be achieved through cooperation, joint development and concurrent engineering during project execution by a variety of partners. Each of them is specialized in his specific field and contributes according to a well-defined and precisely followed workflow scheme.

As already mentioned at the beginning of this paper, it may sound strange to compare the product development process for the mass production of automobiles as for instance used by BMW (see Fig.10) with the plant engineering process as used by Lurgi for a "tailor-made" chemical plant. A comparison done between BMW and Lurgi as part of a benchmark study for the reduction of product development times showed substantial similarities with respect to the methodology used.

In the context of the advanced development of an automobile, individual systems are developed in parallel according to the respective requirements. At the beginning of the actual product development process it will be decided which systems are to be integrated into the new model based on the results of advanced development and by when the corresponding prototypes have to be available.

During the product development process, it is decided at so-called synchronization points whether and which new systems are to be used for the further development process or whether existing systems are to be applied.

At the end of the development process is the prototype of the future vehicle. This is followed by extended test drives, among others, prior to going into serial production and thus handover to the customer.

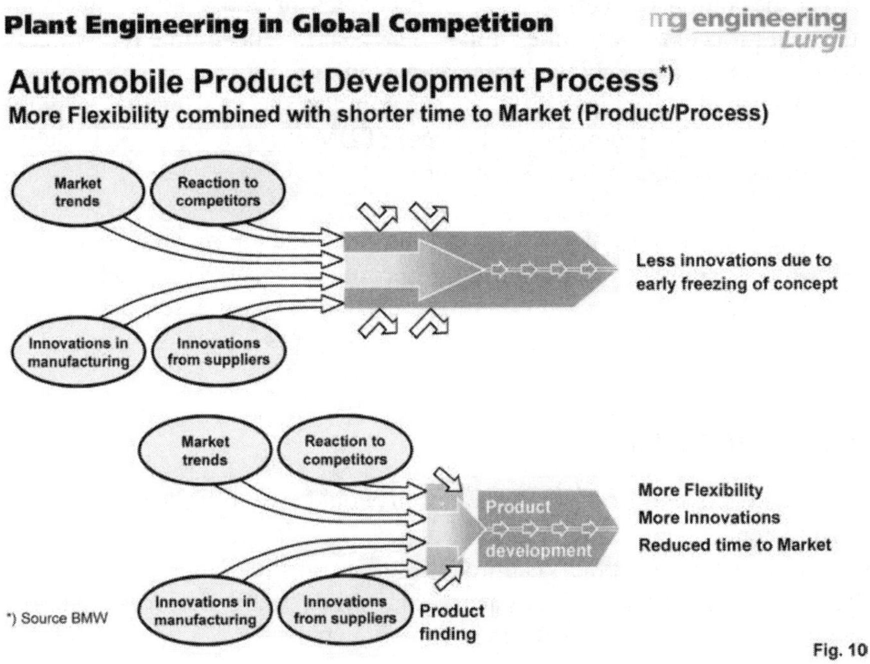

Fig.10: Automobile product development process

Certain similarities exist in the plant engineering process for tailor-made chemical plants. The individual process stages are selected in the context of advanced analyses which are often initiated without direct reference to a project and are followed by new developments or further developments, where applicable, depending on the requirements of the overall process. As the respective plant has to be engineered according to the needs of the individual customer, the exact requirements for the individual process steps will be defined during the so-called basic engineering phase and in more detail during the so-called detail engineering phase.

Therefore, plant-specific advanced developments as in automotive engineering are only possible to a limited extent. Many parameters must be developed in the course of the engineering work either out of phase or concurrently. Just like in the product development process of automobiles, synchronization points are essential to determine the progress and possible impact of deviations on individual process steps and/or the overall process, and to define the corresponding corrective action.

Here too, it must be decided whether new or modified process steps can be used or not. This comparison clearly shows:

The definition of work packages, principles of cooperation and management procedures are quite similar. The goal to reduce time to market is the same and can be achieved using very similar tools.

However, the principal difference remains as the goal of the development process is a "prototype" in the case of BMW and a "real" plant delivered to the customer in the case of Lurgi.

Another common point with series producers also emerged: The plant engineering company is increasingly becoming the "system leader" who integrates clients and partners into the execution of complex plants on a worldwide uniform basis.

Employee Skills

Coming to the end of my presentation, I would like to refer to the future requirements on employee skills.

No matter whether the vision of the future of the company is a global large-scale enterprise or the virtual, networked small business with 16! employees....... the world is becoming a big network.

This may in future also have an impact on plant engineering as we have recently experienced: In a project in England we were suddenly faced with a completely new competitor unknown to us who was awarded the order. It showed later that this was a virtual corporation formed of several small, specialized companies who each contributed their share to the performance of the overall contract. In view of the guarantees required in plant engineering, this will certainly be an exception but it also shows a direction into which our modern DP systems and networked cooperation structures may lead us. This also involves another aspect in terms of the future requirements on employee qualification: The vision of an efficient man/machine system again places man at the center of interest – a great opportunity for the engineer as a source of know-how. More than ever before, the skills of staff are a decisive success factor.

Jaguar Racing – Collaboration in the Workflow?

Steve Nevey

Jaguar Racing

This paper will discuss the business of the Jaguar Racing Formula One team. It will attempt to present it as a medium sized engineering and manufacturing enterprise.

Jaguar Racing competes in the Formula One World Championship, which is contested over a period of around nine months in fourteen different countries and over five continents. Eleven teams compete, each with two racing cars, in sixteen races of duration 300 Kilometres. Each team is responsible, not only for fielding two cars for each race, but for the design and manufacture of those cars. Design, manufacture and development forms a significant part of each team's business. As with many companies in the manufacturing sector, time is of the essence and can benefit enormously by effective collaboration between departments and with external partners.

The History

It all began in 1995, when Jackie Stewart was invited to lay the foundations for a Ford Formula One team. Jackie had enjoyed a long association with the Ford Motor Company, which dated back to the late sixties, when he competed for the Formula One World Championship as a driver, winning the drivers title on three occasions. This time, however, he would enter the Championship, on behalf of Ford, as a constructor/owner. Stewart Grand Prix was established in 1996 and competed in its first race in 1997. Almost immediately, the team enjoyed relative success as a newly formed "small" team, winning its first Grand Prix in 1999.

The plan was always for Ford to have the option of purchasing the team from Jackie Stewart after five years. In fact, it was in 1999 that Ford realised, not only that the team was capable of success, but that it would need to become a larger organisation, with better resources, if it were to compete with the leading teams. Ford bought the team in the summer of 1999, and renamed the team as Jaguar Racing at the beginning of 2000.

The Current Team

Jaguar Racing is currently based in Milton Keynes, in a 127,000 sq ft design and manufacturing facility, and employs around 300 people. The total annual budget for the team is in excess of $170,000,000. For the past three years, however, Sir Jackie Stewart has only had a peripheral interest in the commercial aspect of the business, helping to service a number of key sponsors.

There is a purpose built wind tunnel facility, thirty miles away, in Bicester, where a half-scale model of the car is aerodynamically analysed on a rolling road. The engine is developed by Cosworth Racing, which is also owned by the Ford Motor Company and is based in Northampton, with electronics for data acquisition and control systems being managed by Pi in Cambridge; another Ford owned company. Collectively, Jaguar Racing, Cosworth Racing and Pi are know as PPD, or Premier Performance Division, which is part of Ford's PAG, Premier Automotive Group, which in itself includes Jaguar Cars, Aston Martin, Lincoln, Volvo and Land Rover.

The Technology

There is a variety of technology which contributes the development of a Jaguar Racing Formula One car. When the team was established in 1996, it was decided that only contemporary technology would be employed. The Design Office was equipped with fifteen CAD workstations, primarily running EDS's Unigraphics CADCAM software. That number has increased to over sixty, still without a drawing board in sight! The Unigraphics system is used for three-dimensional CAD (Computer Aided Design) design and CAM (Computer Aided Manufacturing) machine toolpath generation. It is available throughout the company and is easily accessed through the companies PDM (Product Data Management) system. Data can be easily viewed with easy-to-use visualisation tools, removing the requirement for individuals to be skilled in using sophisticated CAD tools to simply see data.

Amongst the other CAE (Computer Aided Engineering) applications employed in the Jaguar Racing Design Office are, FEA (Finite Element Analysis) for both structural and aerodynamic analysis; and kinematics analysis for the full vehicle, which includes engine and transmission loads.

There are a number of physical analysis tools employed at Jaguar Racing, including a half-scale wind tunnel for aerodynamic analysis and a seven post hydraulic rig for full vehicle dynamics analysis. The use of these tools will be discussed in more detail later in this paper.

For Jaguar Racing, like many other manufacturing companies, time is of the essence in the product development process. The aim is to enable the designer to investigate as many iterations of a component or system design as is possible within the limited time available, while collaborating with colleagues in design, manufacturing and production departments to ensure that the finished product is as functionally optimised as possible, can be efficiently produced and is fully appropriate to the end user. For this reason, it is essential that the tools employed are fast, accurate and flexible. Integration between departments, databases and applications is essential to enable engineers, designers and managers to effectively collaborate.

CADCAM with Unigraphics

The Unigraphics CADCAM system provides the team with, not only a comprehensive and state of the art modelling and draughting system, but also a fully integrated NC (Numerical Controlled) machine tool programming system.

Why is this important? By having a fully integrated CAD and CAM system, we can allow the designer more time to continue to optimise component design, even while the NC programmer is generating toolpaths on the designers' CAD geometry. If the designer moves a boss slightly, or modifies the depth of a pocket, the associated toolpaths, in most cases, will update as well. This means that the design and manufacturing activities can take place simultaneously, enabling the manufacturing engineer to start a little earlier and, more importantly, allowing the designer to consider a few more iterations of his design. More iterations mean faster laps.

This also applies to component development during the racing season. On average, there are only two weeks between each race during the season. Every team is continually developing its racing car to improve performance. If a team neglects to develop its car, it will soon find itself sliding down the timing sheet.

Structural Simulation with Finite Element Method

This is an obvious tool for a company in the business of designing and manufacturing a high performance, precision product. Finite element analysis is used not only to predict if a component is strong enough, but also to identify if it is TOO strong. Each component is designed to do a specific job and no more. Why carry 20 grams of titanium around a 2.5-mile race circuit, when it is contributing nothing to the cars performance.

In the words of Sir Jackie Stewart, "We design the car to be ultimately fragile".

Although the car is theoretically stiff, with no bushing other than the tyres, there is compliance in all components, including the monocoque chassis. An understanding of this compliance is vital to the optimisation of suspension design. A small team of structural engineers is able to provide stiffness and compliance data to a similarly small team of vehicle dynamics engineers, before any material is committed to physical prototyping. The aim is to take a fully predictable racing car to its first track test, with no surprises about the way it handles or performs. The only effective way to do this is to deal with analysis holistically. Aero loads from CFD analysis are provided as load cases to stress engineers. Vehicle dynamicists also benefit from aero maps from the Aerodynamics department, to make their models more realistic and, in turn, are able to provide more accurate load cases to structural analysts ... and so on.

The primary FEA tool used at Jaguar Racing is Nastran/Patran from a company called MSC. There is also an ongoing joint project with Altair Engineering, which is aimed at developing a tool for optimisation of composites structures. When the tool is fully developed, the designer will be able to input a three-dimensional shape, e.g. a front wing, along with a set of structural criteria, such as maximum deflection under a given load, and simply let the system calculate the most efficient composite structure. This will free up the designer to use his time more effectively, and will ensure that the design problem has undergone an appropriate number of iterations, prior to committing to manufacture. In fact, this is a four way effort, with contributions from Jaguar Racing, Altair Engineering, Ford Motor Company and the University of Bradford. By carefully setting clear objectives, everybody benefits from this joint project.

Vehicle Dynamics Simulation

Vehicle dynamics are analysed in various different ways throughout the product lifecycle. During the digital definition of the car, vehicle dynamics engineers are able to model and simulate various moving parts of the car using a software product called ADAMS. From the effect of suspension geometry in various cornering, braking and accelerating conditions, to the effect of loads created by the power train; all can be simulated to provide data regarding vehicle kinematics and loads transmitted through suspension and chassis. This chassis load and kinematics data is then used to fine tune suspension geometry and to provide load-cases to the design team and structural engineers to enable accurate simulation of stress distribution. To close the loop, deflection data, e.g. how much the chassis bution. To close the loop, deflection data, e.g. how much the chassis twists, is fed back to the vehicle dynamics department to enable those engineers to build compliance into their simulation model. And remember, all of this is done before a physical prototype is produced.

Aerodynamics Simulation

Jaguar Racing has a team of around 50 people dedicated to all aerodynamic aspects of the car. A general quest to find more downforce, while reducing drag, is balanced with fine analysis to identify the most appropriate and efficient aerodynamic set-up to suit each of the 17 circuits in the Formula One World Championship, from the extreme high downforce configuration of the Monaco Grand Prix to the low downforce and low drag set-up for Monza. In addition, there are aerodynamic requirements to direct air through the radiators to cool the engine; to get air into the engine to facilitate combustion; to direct air through the wheels, to cool the brakes and bearings; and to cool various electronic "boxes" scattered around the car. All of these necessary requirements serve to undo the small percentages of efficiency painstakingly gained in the wind tunnel and with computer simulations.

Until recently, Jaguar Racing had been using a wind tunnel in California. For two weeks out of every month, a team of aerodynamics engineers and model makers would pack up a half scale model of the car, with an assortment of variations on wing, floor, barge-board and bodywork designs, to conduct a programme of wind tunnel testing. Results would be fed back to the UK head-quarters, where another team of engineers would prepare data for manufacturing successful aero options for the full sized racing car, while preparing for the next wind tunnel test in two or three weeks time. Compare this facility with those enjoyed by Jaguar Racing's competitors. Most teams have access to a local wind tunnel for 24 hours out of every day of the month. The top teams of Ferrari, McLaren and Williams have access to TWO wind tunnels. The lack of wind tunnel time has put Jaguar Racing at a distinct disadvantage .... until now.

Three months ago, Jaguar Racing commissioned its own wind tunnel in Bicester, which is only 30 miles from the Milton Keynes facility. This tunnel is available on a full time basis, and is substantially more sophisticated than the facility in California. Already, Jaguar Racing has made significant improvements to the car's performance.

In addition to the extensive wind tunnel testing programme, there are two other means of analysing the aerodynamic efficiency of the car. An obvious one is to analyse the full sized racing car itself. This is done as a matter of standard procedure at any test or race, but is more effective when specific aero tests are conducted at purpose built testing grounds made available to Jaguar Racing by the Ford Motor Company, or sometimes at disused airfields. Various sensors are attached to the car to measure downforce and drag, normally in a straight line. This data is then compared to the wind tunnel results as a validation exercise.

The other, and more contemporary, method used for aerodynamic analysis, is computational fluid dynamics analysis, or CFD. This involves the computer simulation of a fluid flowing over or through a body. In the case of Jaguar Racing, it is air flowing over or through the racing car. The main problem with CFD is that it is very demanding on compute resource. For example, for Jaguar Racing to make a reasonably detailed full car aerodynamic analysis, eight dual processor, UNIX workstations, each with eight giga-bytes of RAM are linked together to solve a single problem over a twenty-four hour period. Each solution yields an aerodynamic data set for the car in a single configuration and at a fixed ride height.

A question which is frequently asked is, "When will CFD computer simulation replace the physical wind tunnel?". The simple answer is, not for some time. Although CFD analysis enables early analysis of aerodynamic ideas, without the need for producing physical prototypes, it is very time consuming, especially for full car simulation. Remember, a full car analysis takes twenty-four hours to complete at a single ride height. To analyse the car at an alternative ride height of, perhaps, fifteen millimetres at the front and twenty five at the back, would take a further twenty-four hours. During a wind tunnel test the car model is suspended from the top of the tunnel and can be manipulated through a whole analogue range of ride heights during a single twenty-minute session. For that reason, CFD is used mainly to analyse local areas of the car, such as radiator or brake cooling efficiencies. This gives the aerodynamics team a head start on new ideas, prior to heading off to the tunnel, making more efficient use of wind tunnel time, thus yielding quicker results.

The aim for the aerodynamics department is to increase the speed and accuracy of CFD analysis, and to obtain 100% correlation between the various methods for analysis.

Rapid Prototyping

Various methods of rapid prototyping have been employed at Jaguar Racing over the years, from SLA (stereo lithography) for components to add to the physical mock-up, which will be discussed in the next section, to wax models for facilitating investment casting.

It is only in the last year, however, that rapid prototyping has really delivered results to the team. Until around a year ago, the only method available to the aerodynamics department, for producing components for the wind tunnel, was to physically manufacture the parts by traditional methods such as machining out of various materials, fabrication, or by composite lay-up. The process would involve the CAD designer producing a three-dimensional virtual model that, although extremely accurate, could afford to be somewhat untidy. In other words, to expedite the process of producing as many iterations of each component as possible, the designer would leave CAD surfaces untrimmed and intersecting with each other, with material thickness implied rather than modelled. It would be left to the model maker to interpolate the design data and produce the physical models as quickly as possible, so as to enable as many iterations of a design idea in the wind tunnel as possible. The bottleneck in this particular scenario was with the physical manufacture of the wind tunnel model components.

The introduction of two SLA machines in the model shop has drastically improved the time it takes to produce components for the wind tunnel model. In fact, dependant on the physical size of each component, several iterations of each design can be produced at the same time. However, it soon became obvious that the bottleneck had not been removed, but had been shifted back up he process chain. The SLA machines rely on a fully closed solid CAD model of a component as input, so the designer can no longer get away with producing his untidy model, but has to complete the trimming of surfaces, etc. Jaguar Racing was still taking the same number of iterations of each design idea to the wind tunnel, because it was taking the designer longer to model each of them, prior to rapid-prototyping.

Jaguar Racing enjoys a very healthy commercial and technical relationship with EDS who provide the Unigraphics CADCAM system to the team. Jaguar Racing enjoys frequent visits from EDS engineers and consultants, who are able to advise on best practices and techniques for using the Unigraphics suite of applications. The SLA modelling problem was described to one of the EDS consultants, who was able to deliver a workshop on parametric CAD modelling. The result of this is that a CAD designer is now able to produce an intelligent model, with a number of changeable key parameters, such as thickness, rake angle, blend radii, etc. The creation of this parametric CAD model takes, on average, four days for the designer to produce. However, the creation of each, subsequent iteration takes only seconds, after tweaking appropriate parameters. In fact, iterations of the design can be created by an aerodynamicist, who is not expert in using the CAD system, but who can modify appropriate parameters through a simple interface and simply ask Unigraphics to update the model.

So now, Jaguar Racing is able to take twenty iterations of a design idea to the wind tunnel for testing, rather than five.

Driver Ergonomics

Up until around five years ago, driver ergonomics would not really be evaluated until the chassis had been substantially defined and a mock-up had be produced, into which the drivers would be invited to sit. The driver would comment on his driving position and, based on his feedback, the design team would go away to re-design aspects of the car. This would often put the development programme back quite substantially, sometimes involving additional wind tunnel testing, if the chassis needed significant modification.

Now, Jaguar Racing employs the help of the Lear Corporation, who scan the drivers and produce fully kinematic, anthropometric models of both of the drivers. The Jaguar Racing design team are then able to fit the drivers into the chassis in the virtual world of CAD. In fact, it would be more accurate to describe the process as designing the car around the drivers. This makes more effective use of the limited design time available, allowing the design team to evaluate more aerodynamic and structural iterations.

Lots of Data

Jaguar Racing exists to produce data. In fact, it would be more accurate to describe the company's end product as being knowledge and intellect. The team is continually generating data in the form of three-dimensional CAD data, engineering analysis results and telemetry data. This data is collected, stored, processed and delivered, to those members of the team who need to see it. If the company's end product is knowledge, the racing car should be thought of as simply a manifestation of that knowledge. The car never stands still, but is continually being developed to improve performance, based on data it generates about itself.

There are in excess of 150 sensors on the car at any given time, collecting data about stresses, strains, pressures, temperatures, wheel speeds, ride heights and more. The data is collected onboard and, to an extent, is processed on board. Some of the data is transmitted, by radio signal, to the team's garage in the pits. The team is also able to transmit data to the car, to fine-tune certain parameters, based on real-time data being received from the car. Other than that, the remainder of the telemetry data is downloaded from the car when it returns to the garage. An umbilical cord is connected to the car, whereby it becomes a node on the network. The downloaded data is then analysed by the race engineers and drivers, enabling them to identify which variables on the car, such as gear ratios, anti-roll bar stiffness, spring rates, wing angles, wheel cambers, etc., can be modified to improve the cars performance and subsequently reduce the driver's lap time.

At the same time, data is being transmitted back to the Jaguar Racing HQ in Milton Keynes, where a team of engineers can also analyse the data to help find those extra hundredths of a second in a lap.

The Seven Post Hydraulic Rig

At the back of the Jaguar Racing factory is a rig room. In that rig room is a full sized formula one car on a seven post hydraulic rig. There is a hydraulic ram under each of the four wheels, two under the engine and one under the nose of the car. The purpose of this rig is to push and pull the car in various directions, for numerous reasons. The rig can be used to validate the compliance results obtained from the various computer simulation tools in the design office. It can also be used to make modal analysis of the car structure, to help the race engineers to identify the best way to apply damping. However, the most interesting application for the seven post rig must be at a race weekend.

The grand prix takes place over 3 days. With the exception of the Monaco Grand Prix, it all starts on a Friday. The teams have two free practice sessions of duration one hour. The teams complete as many laps as they choose. On Saturday morning the teams have two free practice sessions, each of duration forty-five minutes. During these sessions, the teams can complete as many laps as they choose. During these Friday and Saturday morning sessions the teams will learn as much as they can about the track, the weather conditions and the tyre performance, with the primary objective of finding the optimum set-up for the car for qualifying and for the race.

It's on Saturday afternoon that the competition really starts. There is a one-hour qualifying session, when each driver has a total of twelve laps on the track, during which he tries to post the fastest lap he possibly can. The result of this qualifying session determines the grid positions for the start of the race to take place on Sunday afternoon. Each driver will usually complete four sets of three laps, made up of an "out" lap, a "flying" lap and an "in" lap.

On Sunday morning, there is a thirty-minute warm up session, and in the afternoon the twenty-two drivers will race over a distance of approximately 300 kilometres.

To overcome the problem of being limited to only three and a half hours of free practice, Jaguar Racing, like most of the front running teams, uses its seven post rig to simulate the circuit, back at the factory. In basic terms, while the drivers and engineers are asleep in Monza, Melbourne, Indianapolis or Imola, the car is doing laps, on the rig back at the factory, of which-ever grand prix circuit, collecting data which will be translated into knowledge and sent back to the engineers at the circuit, to improve the car's performance for the following day.

Incidentally, the Monaco Grand Prix, unlike the other sixteen, takes place over Thursday, Saturday and Sunday.

Not So Different

One thing that is hopefully becoming evident is that the business of Jaguar Racing is very similar to that of many other, more conventional manufacturing companies. The obvious comparison is with the automotive industry, but the similarities go beyond that. The composites aspect of the business has much in common with the aerospace industry, and the pressures of "time to market" are often recognised by visitors to Jaguar Racing from the mobile communications and telecommunications industries. The FIA will not postpone a grand prix if we are not ready. The fine for not having two cars with which to compete is $500,000.

Just like Ford, BMW, General Motors and Jaguar Cars, Jaguar Racing has various design departments, a production engineering department, a manufacturing department, an inspection department and a stores department. What the others can't boast is that their customer is part of the enterprise. Jaguar Racing is its own customer. Not only is Jaguar Racing a fascinating and concise model of more conventional manufacturing engineering, but it demonstrates the importance of including all those involved in the final product, including the customer, in the product development process. Jaguar Racing produces exactly what the customer wants.

In fact, communication with all aspects of the enterprise, not just the customer, is encouraged at all times. Designers collaborate and share ideas throughout the design process. Production and manufacturing colleagues are involved in the design process, helping the designer to ensure that component and assembly designs are manufacturable and maintainable.

Car performance is never compromised, but cost and efficiency of process are always a primary concern.

Collaboration tools, such as those provided by EDS, make engineering data available to everybody who needs to see it, throughout the organisation. There are no surprises for anybody, at any point. Everyone is aware of every aspect of the design and production process.

The Design Process

The sequence of events starts around August with the Chief Designer providing the design team with a brief. This is based on a set of functional requirements such as length of wheelbase, centre of gravity position, and a set of regulations specified by the sports governing body, the FIA. A set of "regulation boxes" represents an imaginary set of three-dimensional envelopes within which certain aspects of the car, such as front wings or sidepods, must be contained.

Each designer will claim his space in the three-dimensional virtual world. For example, the gearbox design team will model an initial gearbox design, maybe simply a cuboid. The suspension designer will produce a crude model of the rear suspension, maybe simply a cylinder to represent the wheel and some lines to represent the suspension geometry. The important thing is that they are both doing this within the same virtual space, so can see each other's design schemes. This ensures that they are able to develop interfaces between their systems, i.e. the rear suspension is mounted on the gearbox; and to identify and resolve conflicts between their design schemes. This philosophy applies to the entire vehicle, whereby each component and assembly begins as a series of rather crude shapes and gradually takes on more detail, which means that the car gradually "comes into focus" as it is developed.

Manufacturing starts in earnest around the beginning of November. The first iteration of the completed car will be ready shortly after Christmas and is first taken to Silverstone for a "shakedown" test, before being transported to Barcelona or Valencia for more realistic testing in weather conditions more typical of those seen at a Grand Prix.

Design and development does not stop here. More detailed analysis of each component may result in remanufacture in titanium rather than steel, or components may go back onto the milling machine to have fractions of a millimetre shaved off, to reduce weight. Every extra design iteration means fractions of a second off a lap time at the first grand prix in Melbourne.

The first grand prix takes place at the beginning of March. From then on, the grand prix are generally every two weeks. The cycle continues ……… run car; acquire data; generate knowledge; identify design iteration; execute change; run car ……… etc.

Effective Collaboration is Essential

It is absolutely essential that EVERYONE in the organisation collaborates freely. Design and Production departments must be aware of the entire process flow. Production can gear up for manufacturing jobs coming through from design, and design staff can monitor progress through the plant, both from a scheduling and quality point of view. To get the best from suppliers and external partners, the enterprise must be extended beyond the physical and virtual boundaries of the Milton Keynes headquarters. There are certain elements of supplier relationships that rely heavily on trust. For example, a company called Xtrac is responsible for manufacturing Jaguar Racing's gearbox internals. Bear in mind that Xtrac make gearbox internals for most of the other F1 teams. Their business is built largely on trust. Without that trust, they would not do business. The most important element in our process flow is our race and test team …. Our customer!! So now we can see that Jaguar Racing emerges as, not only an interesting and concise model of a more conventional manufacturing company, but an example of how important it is to listen to the customer.

Managing Collaboration

So how do we manage effective collaboration throughout the enterprise? You must secure the full backing of people at the very top of the organisation. An exercise in effective collaboration cannot be approached on a whim. There must be full commitment and support. It is also important, not only to get the full support of senior management, but to get them to identify the key business drivers. An effective, collaborative process can only work if EVERYONE involved understands what the key business objectives are.

Carefully select and then consult with key individuals from each department. Find out what their particular communication requirements are. It is important, however, to ensure that any new process is not compromised by individual requirements. Individuals may be asked to compromise on their particular requirements. If they are, ensure that they are aware that an overall business benefit has been achieved as a result of their compromise.

Although technology will eventually be a key factor in effective collaboration, it should not influence the initially identified process flow. Identify the communication chain and then identify technology to affect it. Do not let your processes be compromised by technology limitations. Choose technology PARTNERS and not just technology SUPPLIERS. Technology partners will bring a wealth of experience, if they have delivered on this sort of project before. They will help you with the most effective deployment of their technology, and will probably be able to introduce you to their own business alliance partners. You do not want somebody who will deliver a software package and then walk away!

Choose your team carefully. A collaborative process flow is best defined and developed by engineers who know your business rather than IT specialists. Your overall project manager should be somebody you know well and who has been involved in your business for some time.

Set clear objectives and maintain a vision for meeting those objectives. Ensure that everybody who is involved is kept informed of progress.

Take the Lead

Don't be afraid to take the lead on a collaboration project. You will be surprised at how many frightened people there are out there. Companies which are afraid to take the first step. If you lead, they will be more than likely to follow. Invite your partners to join a discussion forum. If you can encourage partners and suppliers to join you in developing a process of collaboration, you will find that you can share knowledge, resources and cost. It will also give you added strength for lobbying technology vendors to deliver a more appropriate solution. Why not invite your competitors to a forum? Jaguar Racing does it. We talk to BAR, Williams and Jordan about a lot of common issues, such as HR, processes and IT technology. We do not tell them about our new aerodynamics package, or reveal our latest suspension geometry, but we do share common technology solutions. Individually, we have to protect our businesses, but collectively, we have to protect our market place.

The final message? DO SOMETHING!

The Future for Jaguar Racing

Jaguar Racing is still a relatively young engineering company. So what does the future hold? It has become clear that, if ever we needed a good example of knowledge-based industry, Formula One is it. Compared to its competition, Jaguar Racing has had very little time to acquire the knowledge and experience necessary to win the World Championship. However, we don't want to wait twenty-five years to win the World Championship, like Ferrari did. Therefore, we must develop appropriate technology to gather, process and present data to those who need it. We must enable efficient and accurate methods for predictive engineering, so that the car is as fully optimised as is possible when it competes in the first grand prix of the season. We must ensure that communication is unhindered between those who need to collaborate across the globe and in different time zones.

Jaguar Racing will win the Formula One World Championship, and repay the faith and investment of the Ford Motor Company and Jaguar Racing's numerous other partners, sponsors and supporters.

Long-Term Cooperation in the European Aerospace Industry

A Model for Future Successful Partnering in Competitive Environments

Hans-Erich Mundt

Senior Vice President Corporate Sourcing, EADS – European Aeronautic Defence and Space Company

Compared to many other manufacturing sectors, aerospace industry is characterized by some specific features that have to be taken into account prior to definition and implementation of a group-wide top level sourcing strategy:

- Extremely long product and programme life cycles: more than 30 years of production and in-service phases
- Outstanding technical complexity, e.g. between 2 and 4 million parts to be assembled into one aircraft
- Maximum requirements regarding safety and reliability of our products
- Globalised industrial structure with a highly specialized supply base being in an ongoing consolidation process
- Strong and very long dependencies between cooperating companies in a cyclical economic environment frequently lead to single source relationships exposed to high challenges in terms of cost and risk in case of change of supplier

These typical characteristics necessitate building a specific strategic approach ensuring long-term competitiveness of our supply chain on a partnership basis.

EADS Sourcing Strategy is based on the vision that we wish to deliver competitive advantage to all our end-customer markets by winning, integrating and developing the world's best suppliers in a strategic perspective.

This vision is realized by implementing four major key strategy elements that are the building blocks of EADS Sourcing Strategy:

1. Procurement Marketing:
   - Support of our global sales activities in strategic key regions or countries through integration of the appropriate supply sources in respective areas of interest
   - Analysis of the globally existing capabilities and capacities to reach the best strategic fit and to ensure long-term security of supply interest
   - Market structure and competition analysis up to the creation for new supply sources and / or avoidance of monopolies

2. Procurement Policy:
   - Balanced sharing of global business perspectives in all EADS end-customer markets between our Business Units and the supply base by early involvement in joint product development activities, market specific analysis and sharing of existing sets of opportunities and risks throughout programme life cycles and continuous performance improvement.

3. Supplier Evaluation and Development:
   - Systematic evaluation and development of our key suppliers along group-wide consistent criteria; strategic performance development and monitoring loops according to the targets of our business; for selected high performers joint business development within global EADS market perimeter is offered to assure growth with our best supply sources.

4. Joint Sourcing:
   - Group-wide pooling of common demand across all Business Units and focalised negotiations within the so-called lead buyer concept. The EADS lead buyer coordinates and manages the internal networks of demands and the external relationships to our large multi Business Units suppliers to achieve optimised commercial and industrial results.

On top of the implementation of these key strategy elements we develop and provide to our Business Units sourcing methods, tools and processes to support the implementation of our sourcing strategy in all aspects where common needs are identified. Thus we assure that all procurement departments within EADS can benefit from best practices in those areas that are equal or similar in most of our businesses. This applies in particular to our e-business ventures such as eRFQ's or reverse auctions.

In what we call Strategic Supplier Relationship Management we manage our major suppliers basically along two dimensions:

1. Performance:
   Performance is evaluated and developed according to our common EADS supplier evaluation and development system. We consider the speed of improvement to be more important than absolute performances – the tendency of today's and tomorrow's performances quantifies the particular contribution of each supplier to the success of EADS in our various markets. Suppliers not sufficiently contributing to the overall targets of EADS are not eligible to expand their business within our scope of activities

   And

2. Commitment to EADS:
   Long-term active willingness to co-operate with EADS and share the opportunities and risks of EADS in all our businesses is key to further develop our supply chain in a strategic perspective and work together in all areas of common interest.

The more we can observe positive trends in both dimensions, the more we are ready to offer a given supplier a new and closer type of business relationship to EADS that is called "Competitive Partnership" and leads to further successful business development.

Summary:

Extremely long market cycles of aerospace products require continuous and long-term co-operation between suppliers and customers in a model of full accountability for their respective tasks characterized by

- Early integration of suppliers in the development of future products to focus the best available competences
- Fair sharing of business opportunities and risks in all end-customer markets EADS is active in
- EADS Sourcing Strategy as a lever providing both a group-wide vision and concrete building blocks for company integration in a complex merger environment
- Concept of "Competitive Partnership" pursuing the target to provide and enhance value for all stakeholders including the suppliers.

The Network of Automotive Excellence as a Potential Response to Change in Development / Production and Brand Policy

Wilhelm Becker

Head of the Small Models Product Line Division, BMW AG

A. The automobile industry in a state of change

All over the world, the automobile industry finds itself in a phase in which major changes are taking place. No other industrial sector has changed so fundamentally in recent years as the automobile industry – and will continue to do so.

What are the main causes of this dramatic upheaval?

1. Automobile manufacturers' growth strategies

There have never before been so many segments in the automobile market as there are today, nor so many suppliers attempting to achieve a global presence on this market. Product lifecycles are shortening and the customers' demands for individuality and mobility are increasing.

The automobile manufacturers' growth strategies are leading to a series of changes that will have the following effects:

- the classic added-value structures between manufacturers, outside suppliers and service providers will change
- product programs will become more extensive (full-line providers)
- in the course of globalisation, new production plants will be built and further over-capacity will emerge, so that cost pressure will increase
- product life-cycles will become shorter
- the capital required for new developments will increase disproportionately
- development times will decrease (the crucial term here being "time to market")
- after sales and other services will gain in importance

The BMW Group has also identified the need for change and is pursuing a clearly defined strategy, namely concentration on the premium segments of the automobile market. In the next five to six years, BMW will launch 20 new models – a threefold increase in the number of model lines compared with ten years ago.

The Figure 1 illustrates BMW's growth strategy:

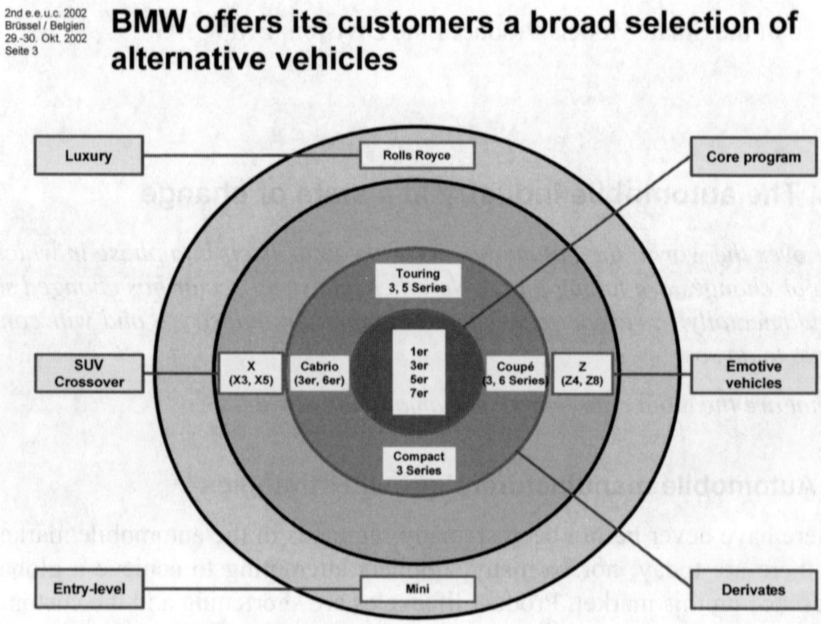

Fig.1: BMW's growth strategy

2. Changes to overall conditions

A series of changes is taking place in the automobile industry and is fundamentally altering the overall conditions that apply to automobile manufacturers.

• More stringent requirements imposed by customers

Customers' individual requirements and expectations are changing. Whereas it was previously enough to offer the customer an affordable means of transport, reliability and a healthy price-performance ratio are today regarded as self-evident. To an increasing extent purchase decisions are based on characteristics that only have a remote connection with the actual transportation function. They include the "driving experience", quality, the vehicle's comfort and convenience, its safety, the communication media and telematics that are provided, all of which now influence the customer's purchase decision.

• Increased mobility requirement

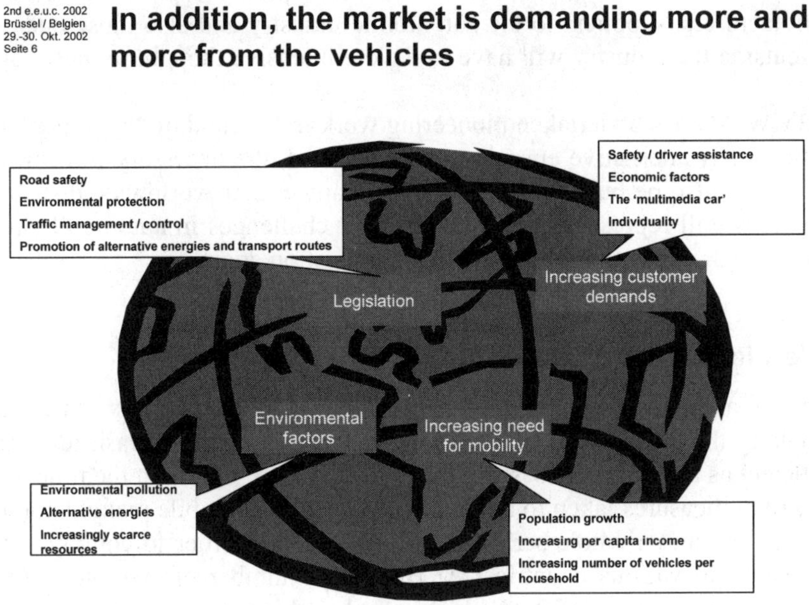

Fig.2: Increasing demands

Until the year 2025, according to a survey conducted by JSAW (Japanese Society of Automobile Engineers), the population will continue to grow rapidly.

The current world vehicle population is already estimated at more than 2 billion. If the number of vehicles rises in proportion to the population density, industry will be obliged to produce increasing numbers of vehicles in order to satisfy demand.

• Shortage of resources, environmental pollution

The automobile industry must also concern itself increasingly with the topic of environmental protection. This is necessary first of all because living-quality demands call for the use of low-pollution vehicles, and second because petroleum as a resource will become increasingly scarce. This will lead to the price of oil rising drastically and a marked increase in the significance of alternative energies.

Whether the answer lies in hydrogen, natural gas (CNG/LNG), hybrid vehicles or fuel cells, the research and development departments of the manufacturers, suppliers to the automobile industry and also those operating outside the industry will have to devote intensive thought to these topics.

BMW AG has undertaken pioneering work in the field of "hydrogen" in particular. If alternative energies are to be used, the necessary infrastructures must also be built up in order to guarantee their worldwide distribution. This will represent one of the greatest challenges in automobile history since the invention of the internal combustion engine.

• New legislation

The increasingly stringent requirements imposed by law are best illustrated by taking the "crash test" as an example. In 1972 a single crash test was sufficient as a means of assessing the risks that were known at the time and the safety measures taken to avoid them. Today, automobile manufacturers have to conduct a whole series of different tests in order to comply with legislation in various countries. At BMW the number of crash tests that have to be carried out is 39, to which must be added an even larger number of static and dynamic tests.

It is in the automobile manufacturers' interest for a high level of harmonization to prevail in the "triad" countries, since this would simplify the complex approval procedures and reduce the cost and the amount of time devoted to development work.

Manufacturers that sell their products worldwide in particular are obliged to incur disproportionate expense in order to satisfy the legal requirements on individual markets.

B. The need for new forms of cooperation

1. New tasks facing automobile manufacturers

The new challenges go beyond the automobile manufacturers' installed capacities, on which the reduced development times and shorter model life-cycles in any case impose an additional burden.

The automobile manufacturers will be unable to master these many and varied problems alone. so that inter-company cooperation will be essential in the future.

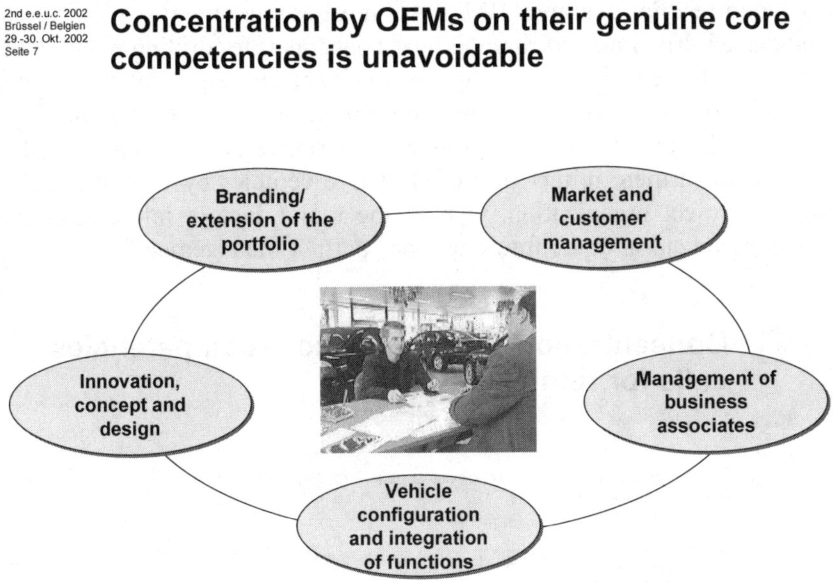

Fig.3: Core competencies

During the 1960s, the automobile industry had almost all the development and manufacturing processes needed to build its vehicles at its disposal "in-house". In future, however, automobile manufacturers will be forced to concentrate increasingly on their own genuine core competencies and to collaborate with and integrate their business associates into the product creation process (PEP).

Vehicle manufacturers will increasingly shift the emphasis of their activities on to the tasks and services that have a direct relationship with the customer. Customer contacts and the associated information will enable the product's properties, after-sales service, other services and all forms of link with the customer to be implemented in a manner that promises success. Together with branding, this will permit a decisive competitive advantage to be achieved.

2. The new demands and tasks within the network

Automobile manufacturers must none the less make sure that they retain their identities. For example, BMW AG must attract attention by virtue of its advanced driveline and running gear concepts (the 'driving experience' it offers), advanced design and modern control and ergonomic concepts. This calls for extensive innovation, configuration and integration management in order to establish the product's substance and its characteristics and implement them in the form of attractive vehicles by way of a highly efficient project organization. One of the major factors influencing this project organization in the future will be "partner management".

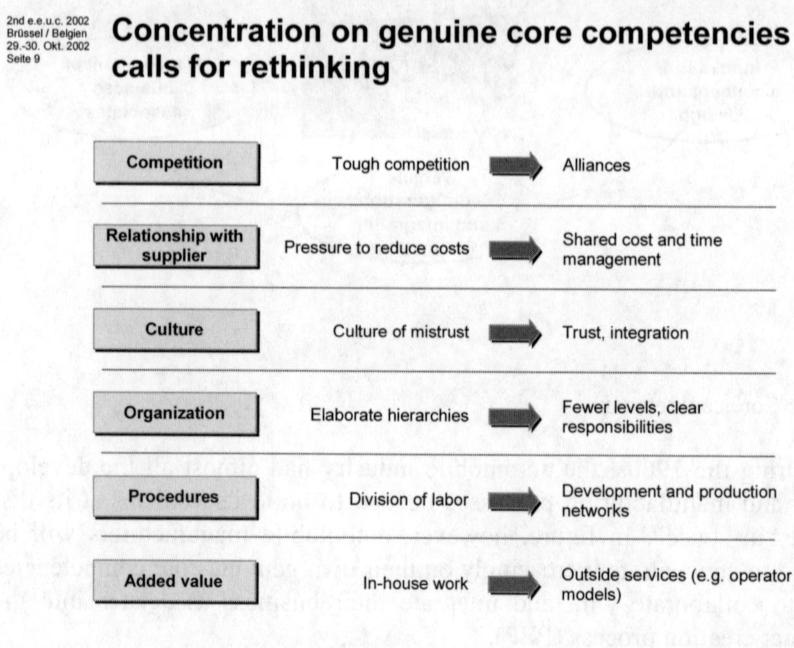

Fig.4: Core competencies and rethinking

• Progress = rethinking within the company

Whereas in the 1990s we were mainly concerned with process improvements ('kanban', lean production, JIT, 'kaizen', re-engineering), the next decade will give prominence to the question of collaboration and networking (integration and communication).

Rethinking in the face of the changeover to collaboration and networking affects every aspect of corporate activity:

- 'Lone fighters' are superseded by efficiently operating alliances
- Shared cost and time management and process synchronization take the place of sequential processing and cost pressure (project orientation)
- Promotion of trust and integration of associates into the added-value structure at an early stage take the place of a culture of mistrust (in the words of Professor Milberg, "Working alone means addition, co-operating with others means multiplication")

3. Collaborative work

Organizations with a simplified hierarchy, use of electronic technologies, clearly defined responsibilities and target agreement processes permit more rapid decisions and achieve a high degree of transparency compared with more complex management and decision-taking processes involving a series of hierarchical levels.

Development and production networks take the place of classis division-of-labor procedures. In this connection it is important to find the optimum partner in each case and for both internal and external networking to take place.

New collaborative models grow up (for instance operator models) and supersede in-house manufacturing processes that involve higher investment, are cost-intensive and do not exert the necessary differentiation effect.

The relationship between the vehicle manufacturer and the supplier will also change dramatically.

The concept of network organization as it is being discussed today is not fundamentally new, since there were initial attempts to create networks in past decades as well:

- Examples that have become legend are the development and production of the Beetle Convertible or the BMW M1.

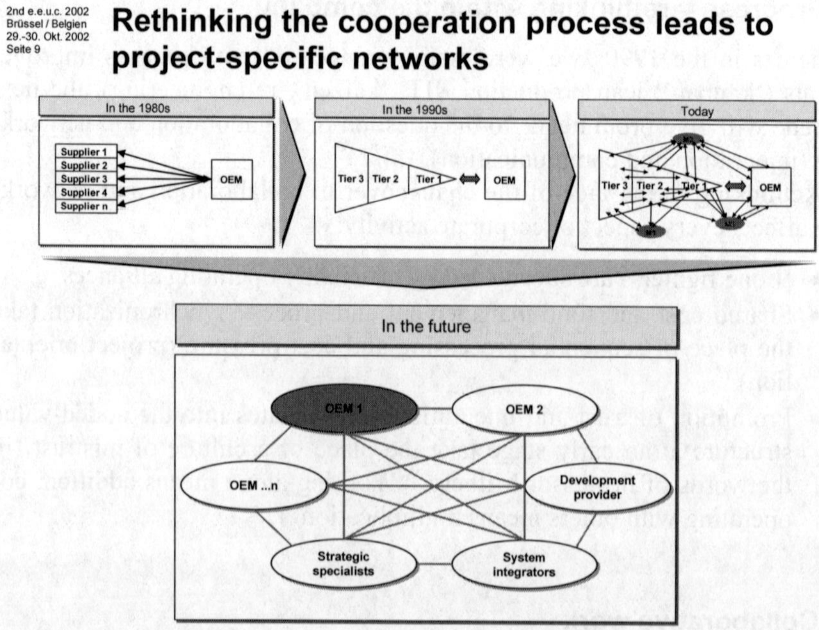

Fig.5: Project-specific networks

- Valmet acts solely as an assembly partner for the Porsche Boxster (a means of satisfying peak demand)
- CKD vehicle assembly in closely regulated markets in Asia, Africa and Central and South America (local content)
- Formation of networks among competitors, for instance engine deliveries to PSA/Ford, BMW/Opel or VW/Volvo
- Joint ventures such as BMW/DC (Tritec, Brazil, engines for the Mini), ZF/Ford (CVT)

In each of these examples cooperation was decided upon in order to satisfy specific demands relating to time, cost, know-how etc. that could not have been coped with so successfully, if at all, alone. The reason for this is that the demands that automobile manufacturers have to satisfy are now so high that only permanent cooperation in the form of networks can solve them.

C. Network of Automotive Excellence

1. Cooperation within the network

According to a frequently encountered definition, corporate networks are structures in which companies work together to achieve shared higher-level objectives. Both the responsibilities and the benefits of these networks are distributed among the partners, so that there is a joint interest in achieving overall success.

Networks are formed in connection with specific projects, in other words as a means of achieving a mutual objective, and wound up after this objective has been realized. Despite the time limit which this imposes on the cooperation, it is ideal if the partners maintain a long-term relationship. This can be expected if a healthy state of equilibrium exists between their contributions, that is to say if a benefit-sharing model is adopted.

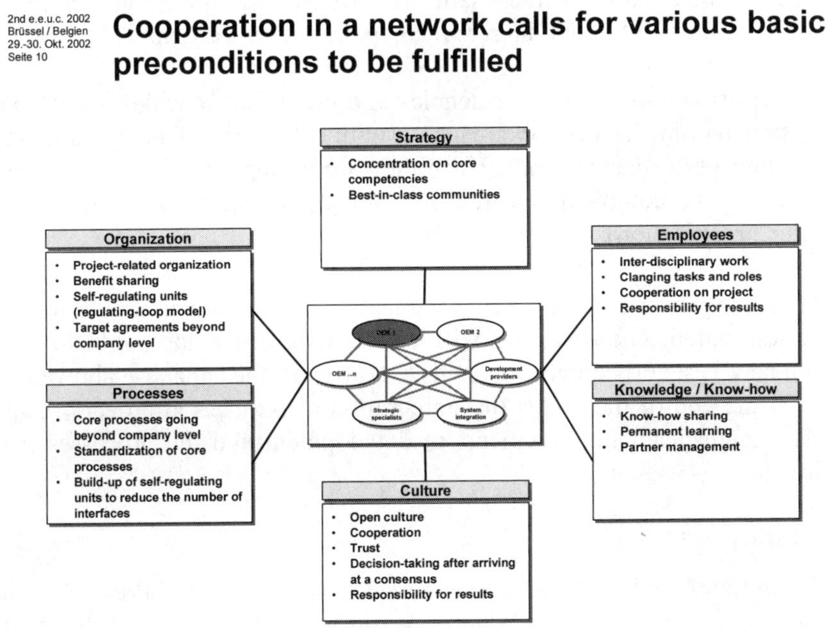

Fig.6: Cooperation in a network

The need to concentrate on the "genuine" core competencies calls for a strategy of best-in-class communities or networks. If companies are to work successfully within the network on a long-term basis, they must create the necessary preconditions. Organization, procedures and employees must be geared or aligned to work within the network, and the necessary knowledge for configuration and management of the networks must be built up.

2. Precondition for a Network of Automotive Excellence

In order to ensure that the network organization is a long-term success, the network concept must be established within the company by a determined "cultural change".

Two aspects of this will be examined here in more detail:

• Organization

Companies have traditionally been units operating autonomously and organizing their inputs in such a way that the desired output materialized at the end of their added value chain. The well-proven principles of functional division of labor and hierarchical coordination were upheld.

Concentration on core competencies and the need for wide-ranging cooperation within the networks cause traditional forms of organization to reach their performance limits, since they are no longer capable of coping with increasing complexity within the company (project organization versus line organization).

The changeover to a network organization is a task that should not be under-estimated. The first steps will lead to disturbance and confusion until the new system becomes instinctive. However, the Japanese showed us back at the end of the 1980s that integration of partners at an early stage leads to significant improvements in development and the processes that follow it.

• Employees

When the transition to networks is made, employees find themselves in a zone of tension, a kind of "double world". One the one hand the traditional hierarchical linear structures still exist, but on the other hand the employees become conscious of the power of the network structures in their day-to-day work. Two entirely different methods of rational thought thus collide, but the employees are expected to find their way about successfully in both of them.

Networks need to be managed. Employees' qualifications must grow as the new tasks arise ("network managers". Leadership becomes more challenging in a network environment, and more is demanded of the employees concerned.

More complex technical methods and sequences call for people capable of thinking in multifunctional terms and assuming responsibility for their actions. The comprehensive know-how, creativity and personal initiative of globally active employees are the keys to the success of the network.

3. Network of Automotive Excellence (NoAE)

The network organization alongside the organization of the OEMs and also those of the outside suppliers and service providers.

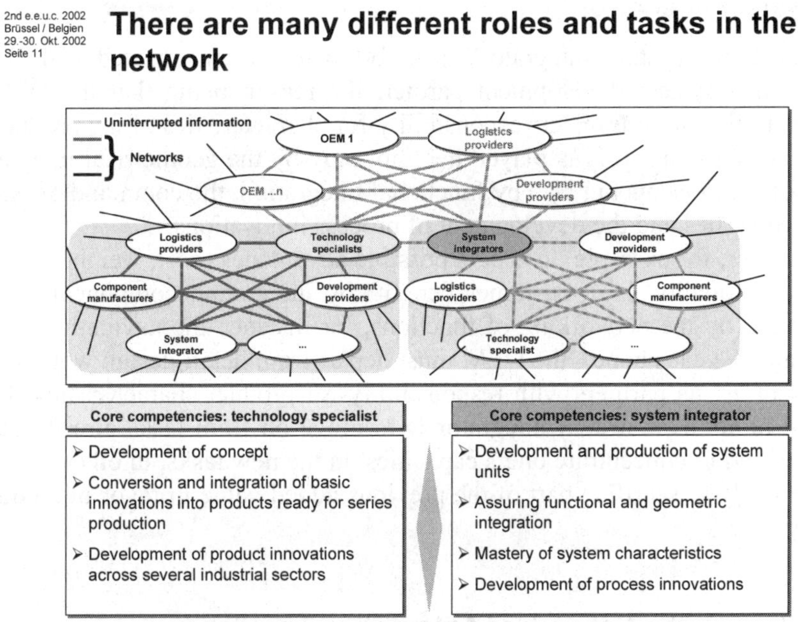

Fig.7: Roles and tasks in a network

Concentration on core competencies and the best-in-class principle lead to the strengthening and modification of the modularisation process in the added value chain. This gives rise to networks built up according to the modularisation principle.

In this context, the automobile industry's central partners are the technology specialist and the system integrator. Both of these partners exert a decisive influence on the product's success.

- **Technology specialist**

In the new network the technology specialist is the innovator who creates products ready for series production from basic innovations and bears the responsibility for the successful integration of the technology into the vehicle. The vehicle's degree of technological innovation stands or falls with the performance of this function.

Technology specialists can achieve cost benefits by developing product innovations or solutions that are not limited to a specific industrial sector.

- **System integrator**

The term "system integrator" is unfortunately often equated with the tasks of a system development partner, the reason being that the OEM finds it difficult to transfer responsibility for characteristics of the product. However, a major role is played here not only by the geometrical integration of components but also by functional integration, the command of system properties and the development of process innovations.

In order, for example, to access possible advantages (cost, weight, space required, quality, high customer benefit) by integrating components and function (by the networking of functions, for instance the navigation system and the car phone), the OEM must depart from detail design work and also entrust its partners with responsibility for product characteristics. To claim to act as a system integrator is to move on from tasks already accepted and to concentrate one's capacities on the new tasks, in other words to be ready to sacrifice part of the previous added value in favor of modularity.

4. Shaping the Network of Automotive Excellence

In addition to the development and production of parts and components, services such as supplier management and therefore complexity management, quality management, finance management and sequencing are now included as well. The term "supply chain network" refers to this new network of evolutionary business links and modified task allocations within the product creation process. Second-level suppliers in particular are no longer directly linked with the OEM but are part of the network of business relationships. This approach enables the company's energies to be concentrated on the most important processes.

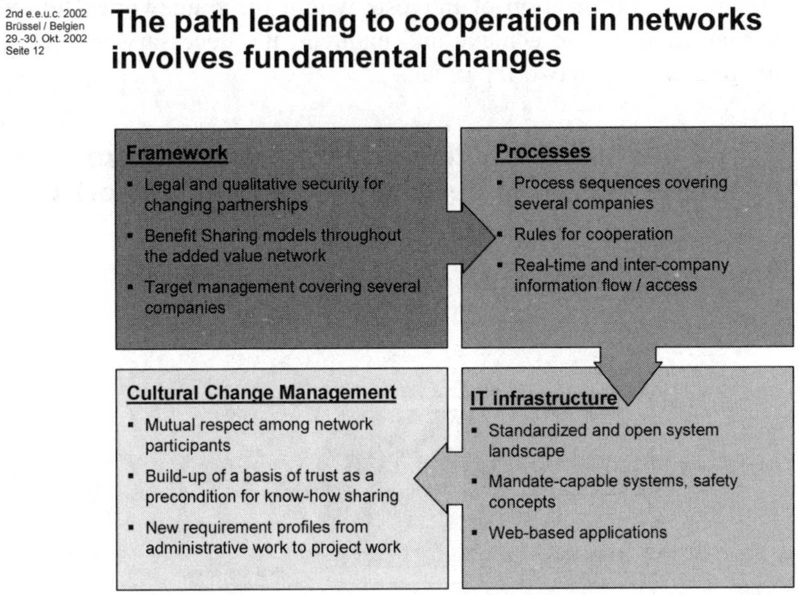

Fig.8: Fundamental changes

In order to handle the increasing complexity, the network partners must link their processes more and more closely together. However, few companies have so far adapted their methods genuinely to permit network cooperation. "Most companies have neither a strategy nor network-capable business processes.

The aim must therefore be to create the preconditions for a successful future using networks.

Network organization needs a framework that can ensure legal and qualitative security in such a way that all participants benefit from the cooperation. This also includes an equitable distribution of economic benefits via the added-value chain.

Management functions such as inter-company target management, monitoring of rules and the control of escalation mechanisms in the event of deviations will soon be part of standard knowledge in companies.

Since in the future companies will be active in several networks and perform different roles within them, they need standards and clear directives n order to monitor the necessary coordination effort. New electronic technologies permit entirely forms of communication and information exchanged to be practiced in this area.

By ensuring the protection of interests within the framework, Cultural Change Management can achieve and maintain the necessary mutual respect and trust among network partners.

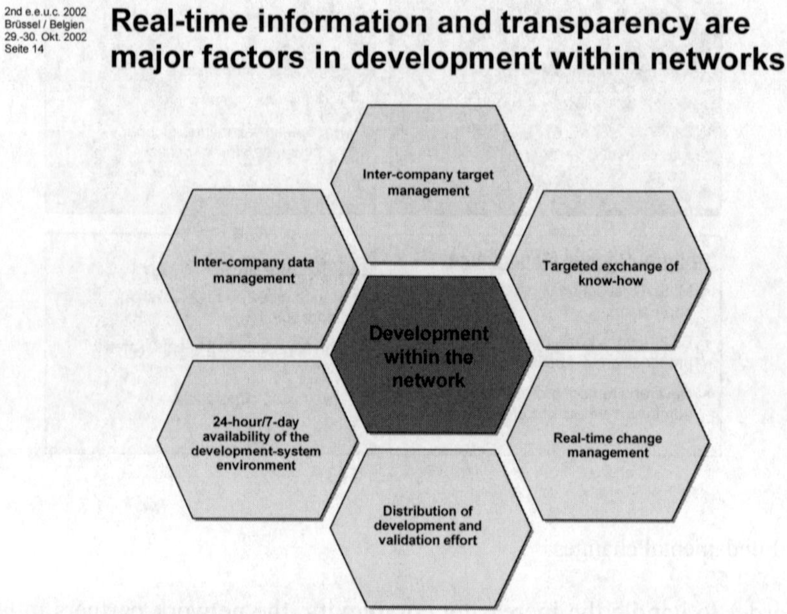

Fig.9: Information and transparency in networks

The vehicle manufacturer's internal development departments already perform coordination tasks to a great extent. None the less, re-alignment within the development area too will be unavoidable in view of forthcoming challenges.

Acceptance of the best-in-class principle by vehicle manufacturers will further reduce the volume of development work that they undertake. The higher proportion of development work will take place at the best-in-class partners within the network, with innovations and services that serve as product differentiation. The manufacturers' development departments will concentrate increasingly on tasks such as establishing objectives, target management, budget monitoring and the necessary coordination.

The performance of development work at more than one location becomes possible as a result of a combination of virtual product development tools and electronic technologies. Real-time information provides an overview of current targets, modifications and the actual state of development.

The increasing variety of models and the reduced development times and product cycles increase the need for concentration. The challenge is to reduce the effort and expense of coordination significantly by adopting electronic technologies, so that no additional expenditure on staff for coordinating functions is needed.

5. Example: the future role of the OEM purchaser in the Network of Automotive Excellence

Considerable changes will also occur in the OEMs' purchasing departments. They will have to limit their influence when choosing sub-suppliers to the work of establishing these suppliers' characteristics, or finalize with the partner at an early stage, which components are to be adopted from the OEM's relevant modular program.

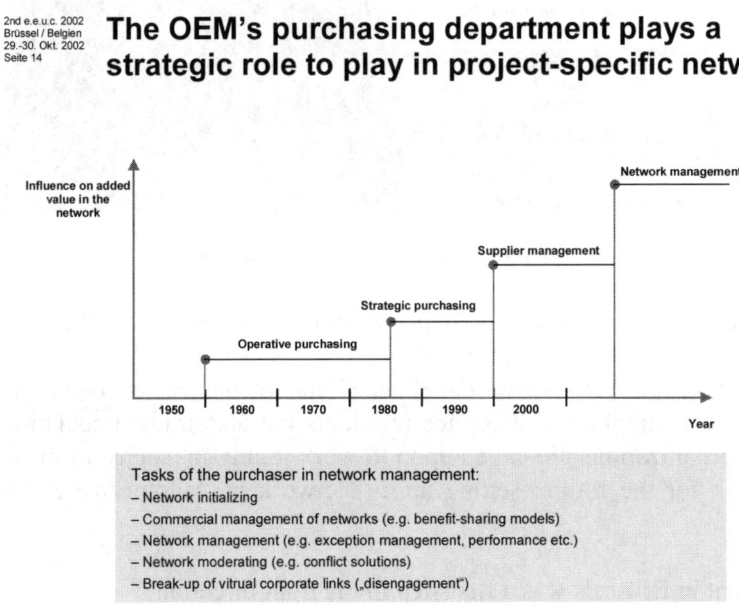

Fig.10: Role of purchasing department

The new tasks of the OEMs' purchasing departments will relate to configuration, management and dissolution of the networks.

D. Outlook

Fig.11: Network of Automotive Excellence as a joint initiative

Future tasks in the automobile industry can only be solved jointly.

A "Network of Automotive Excellence" cannot be created overnight. Manufacturers, suppliers and service providers but also trade associations and other organizations are called upon to work jointly in a spirit of dialog on this task for the future: setting up a "Network of Automotive Excellence".

The event in Brussels was a first step in the right direction.

The European Automotive Supply Industry at the Crossroads

Peter Stehle

Freudenberg & Co. KG, D-69465 Weinheim

1 The automobile industry in the year 2002

From an economic point of view, the year 2002 was, not only as a consequence of the events of September 11th, a hard year for the European industry. This was also true for the automobile industry. Fears that the seven prosperous years from 1995 to 2002 could be followed by seven lean years are frequently being voiced. Others look to the future even more pessimistically and are already asking themselves whether a recession is inevitable.

One can oppose this question quite clearly. The automobile industry market is one of the largest homogeneous markets worldwide and, despite the uncertain economic situation, a stable market. Nevertheless automobile manufacturers are under pressure. Pressure which results from changing circumstances and consequently makes new demands on automobile manufacturers. The tendency towards globalisation has brought increasing international competition in Europe as well. This competition forces one to bring new models onto the market in ever shorter periods of time in order to offer the customers attractive offers. But the shorter runs of these models also lead to shorter development periods, while, at the same time, an increasing number of electronic parts in the vehicles lead to a higher complexity in the development. In addition to this, the main growth in the unit numbers at the big automobile manufacturers no longer comes from the mass produced models but ever more from niche models whose development ties up additional resources. But not only the market, but also the legislator, by means of stricter environmental regulations and expanded consumer protection rules, add to the changed circumstances with which the automobile industry has to deal.

Currency fluctuations and in part short-term changing political circumstances are also increasing the pressure on automobile manufacturers. They, however, are not the only ones who need to adjust.

2 The situation for a automotive supplier in the year 2003

The compulsion for change affecting the automobile manufacturers is passed on to the system suppliers and from these to the component suppliers. The latter have to contend with ever lower prices, higher investment output, increasing risks with recall actions and the assumption of responsibility for technical innovations and development output. Dramatic changes occur. The automobile supply industry is facing structural changes and must adjust quickly and must do this differently from in the past.

The strategy of automobile manufacturers of giving entire modules from development until assembly to automobile suppliers and of therefore reducing their own complexity (and also saving costs, as the own development of many modules is hard to justify economically) has, in latter years, led to the formation of so-called mega-suppliers who are also able to offer these complete modules. These mega-suppliers, frequently also called Tier 1 in the supplier pyramid have also, as a consequence of their size and expanded supplies, caused an increase in the internal complexity. Mentioned can here be made of new demands such as Simultaneous Engineering, global product management, E-Business, Web Based Services and the software for many electronic systems. Simultaneously, the external complexity between automobile manufacturers, systems and component suppliers increases constantly. Time-to-Market demands are becoming shorter and shorter, rising demands on quality, flexibility, project management and the upholding of international laws, as well as global fidelity to delivery dates and ability to compete on a world scale, are accompanied by increasing tasks and costs for development, guarantee performance and the covering of risks. With that it becomes essential to successfully overcome the impetus of the process of change.

In order to reduce this external complexity, automobile manufacturers not only wish for fewer suppliers but also for better suppliers. With a lower number of suppliers comes also a reduced effort of control- but only if the processes are arranged in such a way that, in accordance, the number of contacts are reduced and the suppliers, due to the increased internal complexity imposed on them, do not require additional support on the part of the automobile manufacturer.

From the enterprise point of view of the automobile supply industry on the other hand, the increased complexity means that they, in order to increase their efficiency, must simplify themselves. They must focus themselves. The system supplier with his higher demands is subjected to a particular change. He must rely even more on qualified suppliers. Here, the technology specialist is needed. A technology specialist bases his long-term focus on technology. With his highly specialized knowledge he solves the problems of the customers from the application in the development phase up to the spare part replacement phase and is responsible for supplying customers everywhere in the world. As the focus is based on the technology and not on the application and this often finds a use outside of the automobile industry in general industry, considerable potential for synergy arises which in turn is of benefit to all the customers of the technology specialist.

3 Common goals of automobile manufacturers and automotive supply industry

Despite the different routes which the automobile manufacturers and the automobile supply industry take in order to meet the present demands, both groups are aiming at the same goals. These can be reduced to one slogan: quicker, better, cheaper. Both groups have recognized that there is an advantage in driving forward common goals through constructive dialogue. The model of the arrangement of such a dialogue is to be found in Japan. Worldwide, the development of the Japanese automobile manufacturer Toyota Motor Corporation is, in its continual development, the benchmark for quality, strong balance sheet structure and cooperative collaboration with suppliers. Although in 1954 Toyota motors was still a relatively small automobile industry enterprise of perhaps local importance in Asia, it already operated on the basis of a strategy of participation in technologically orientated enterprises also outside of its *keiretsu,* at an early stage, and now belongs to one of the most successful car manufacturers worldwide. Its participation allowed it to exercise a certain influence on the supply enterprises without however making it dependent on just one enterprise. At the same time, one was proud of the fact that competitors like Mitsubishi, Nissan, Honda or Mazda participated in the same enterprise, as this, on one hand, demonstrated the technological strength of these supply enterprises and, on the other hand, also made sure that the respective technical and most economic solution, independently of who had originally promoted it, found quick admission into its own products. The example of the first hybrid vehicle built in series by Toyota, the model Prius, impressively shows that this strategy still works today. The network of participations led to a company like the Nippon Oilseal Kyogo (NOK) not only being the partner in the sealing technique for today's technologies, but also for tomorrow's technologies, as, for example, it is needed for hydrogen fuel cells. In Europe however, the strategy of securing progress for oneself in joint effort with the suppliers and not granting competitors access to these technologies has till now been predominantly employed. This isolation, however, can no longer be a model for the automobile industry in the future. It must, in order to stay ahead, create modern networks, which are orientated towards the example of Japan but which do not imitate it, but instead, adapted to European conditions, lead to a new level in the industrial society.

4 How can European automobile suppliers preserve and improve their competitive ability?

As the main problem of both the manufacturers and the suppliers lies, as already cited, in the highly increased and still increasing complexity and both groups are aiming for the same goals ("quicker, better, cheaper"), the thing to do would seem to be to reach these goals together by means of joint measures: The measures which need to be taken can be divided into four groups:

- simplification and standardization of processes
- focus on long-term partnerships
- focus on core competences
- permanent improvement in efficiency

These measures can, corresponding to today's requirements, not be realized in statistical but only in real life, constantly developing networks whose foundations are collaboration, mutual confidence and trust.

4.1 Simplification of processes

A simplification of the processes for both the automobile manufacturers and the big tier 1 suppliers can be achieved by, among other things, a reduction in the number of suppliers. This leads to a lesser coordination effort, dramatically reduced development times and a smaller amount of development runs. Simultaneously the process promotes globalisation, as only one supplier comes into question who can also ensure on-site production in the three most important continents for the automobile industry, Europe, America and Asia, as one would otherwise need additional suppliers for these continents, which would oppose a reduction of suppliers. But the reduction in the number of suppliers leads to further simplifications. No supplier will use different production methods for his products but will instead standardize these. This also requires standardized materials and therefore reduces the effort involved in qualification quite considerably. At the same time lower costs can be achieved by using scale effects connected with larger quantities. The investment volumes sink, as more identical parts can be used. A further possibility of simplifying the processes is to reduce the interfaces. Flexibility can be won by the formation of small autarkic highly competent units. This flexibility however is only really effective in networks which embrace many enterprises.

4.2 Focus on long-term partnerships

As already shown using the example of Toyota's suppliers, the most effective way of building up knowledge in order to become a technology leader or to remain one is to exchange it regularly with customers. This only works in long-term collaboration. If one were only to work together for a short period of time, concerns would quite rightly arise that knowledge was perhaps being sucked out of one side and being passed on to competitors. A long-term collaboration must be marked by mutual confidence, trust, the will to actively contribute innovations oneself and the permanent improvement of ones own efficiency. This will bring about a focus on the core competences in the enterprise, which will be linked to the giving up of tasks and business fields which do not belong to the core competences. As, however, in the products of tomorrow many fields of competence will be involved which will then no longer exist in every enterprise unit anymore or never did exist there, it is expedient to collaborate in networks which are all-embracing with regard to enterprises. Similar to the way things are already done internally in enterprises in the context of project work, it will, in a further step, become necessary to work together with alternating partners. This by no means represents a contradiction of long-term partnerships but merely emphasizes the importance of formulating clear and reliable rules of play between partners. In an atmosphere of mutual confidence it will become possible to build up trust. The open collaboration between the suppliers and automobile manufacturers will automatically lead to fewer, but more competent suppliers and they will build up development centres close to customer development. The exchange of technology between the individual development centres of the supplier will propel the best-in-class principle and contribute to quality leadership. An also indispensable costs leadership can then be realized by means of a modern global production concept, as, for example, represented by a combination of lead centres, partner productions and global partner productions. But it is also necessary for trust to be the basis of a long-term partnership for another reason. A reduction of suppliers to one supplier who globally becomes the only partner of the automobile manufacturer for the development, production and part-replacement phase of a module or a technology is only possible and favourable if the supplier is a stable partner who can be relied on in the long-term und who will still exist in fifteen years. Thus, there will be a further consolidation of the numbers of suppliers in the medium-term, as many of today's suppliers have neither the organizational nor the financial power which they need to take the first step into becoming global on their own strength.

4.3 Focus on core competences

Focusing on one's own one core competences means a further step in the reduction of complexity. This however requires one to trust a partner to so that that the partner already counts the competences which have been given up among his core competences or, for example, when hiving off an enterprise, makes them into one of his core competences. On the other hand, such a step also requires the courage to believe in one's own capabilities in one's core competence area and to constantly develop one's own core competences further.

If one takes the example of the hydrogen fuel cell, in the matter of which the experts are all agreed that it will be used in great quantities in the automobile drive in about 20 years. But in order to participate successfully in the business with the hydrogen fuel cell seals in 20 years, an enterprise of the seals branch must already today be integrated in the sealing problem of these drives and be working on the solution of tasks such as life expectancy, media compatibility and security of these seals. Might such a thesis not be very popular at a time in which the "fast followers" instead of the first movers get the deals, one must not forget that these are long-term deals which are not comparable with those of information technology or telecommunications. Naturally it is not enough to believe in one's own capabilities and to further develop core competences, if efficiency is not also further developed and increased in parallel. The demand here is persistence until production finally takes place. If an automobile supplier is successful on this count, he does not have to worry about the fast followers among the competitors.

Even though many European automobile suppliers even today still lay the main focus on either product innovation or process innovation, the future will soon belong to those who improve the products and the process automation equally. No revolutions are asked for which are intended to lead to improvements in leaps and bounds and then get bogged down at the planning stage but rather constant improvement in feasible steps which do not overtax the organizations, can be realized quickly and whose results can be seen in the short-term.

4.4 Permanent efficiency improvement

Potential for improvement in efficiency is not only to be found in the reduction of complexity in the teamwork between automobile suppliers and their suppliers. Rather, potential can be found in the entire supply chain if attention is paid to customer use consistently enough. Materials contribute to improvement in efficiency just as much as the use of economical worldwide production sites. Economical does not automatically have to mean production in low-wage countries, as this can become expensive very quickly if the logistical problems of the cost of worldwide just-in-time supplying are not solved together with the automobile manufacturers. Global service means more than global supplying of products; in the case of possible problems, car manufacturers wish to have the supplier's expert on-site quickly. The most important improvement in efficiency however is to be seen in the role of the quality leader. The customer of the automobile suppliers is the automobile manufacturer and defects in quality do not only cause organizational effort and expenditure for the automobile manufacturer which are then passed on to the supplier but also dissatisfaction among customers, the car buyers, when defects or even failures occur in the vehicle. The automobile supplier is not only interested in the position of the quality leader because of the preservation of a long-term partnership with the car manufacturers but also because of the successes which can be assessed in monetary terms, especially when payments of guarantees are demanded.

The resulting use of well-known tools such as Six Sigma, the proactive risk management of quality and continual improvement of process achieved, for example by the Growth® ("Get rid of waste through team harmony") culture, lived throughout the Freudenberg group, are essential for permanent improvement in efficiency. The goals of proactive risk management of quality are fewer recall actions and a lower risk in quality with simultaneously more variants. A PPM-rate of nought is strived for.

5 Conclusion

Because of the currently developing structural change in the automobile industry, the European automobile suppliers stand at the crossroads. They could, if they do not adjust to the new conditions quickly enough and follow new paths, lose their leading role in the world. On the other hand, the structural change also offers the chance to use this period of upheaval and to preserve or even to increase the potential for competition.

This time, the goal is not only to become bigger but also better and to consequently focus on customer use. Becoming quicker, better and cheaper is, in the long run, not going to be possible by going alone but is only possible together with all enterprises involved with the automobile. To this end the deciding step in the setting-up of networks which embrace all enterprises must be made in order to preserve the lead in technology in the long-term and to secure the jobs in the European automobile supply industry through more responsibility. In order to construct these flexible networks one needs to rethink previous collaboration. Only if it results in partnerships which are reliable in the long-term can the daily increasing tasks be solved more quickly, more efficiently and with a controllable risk of quality. As of yet, these networks do not exist. The "Network of Automotive Excellence" could be the first to accompany the European automobile industry on its way into the future. Let us start this initiative together as partners now!

Network of Automotive Excellence –

The Future Role of *Tier 1* from a Development Standpoint

Peter Köpf

ZF Friedrichshafen AG

The structures within the automotive industry have changed. For instance, the number of independent vehicle manufacturers has fallen from 36 (in the year 1970) to 11 (in the year 2001; Fig.1). At the same time, there has been a sharp increase in the number of vehicle variants – i.e. vehicle models, engine versions and trim levels – coupled with substantially shorter product lifecycles (Fig.2).

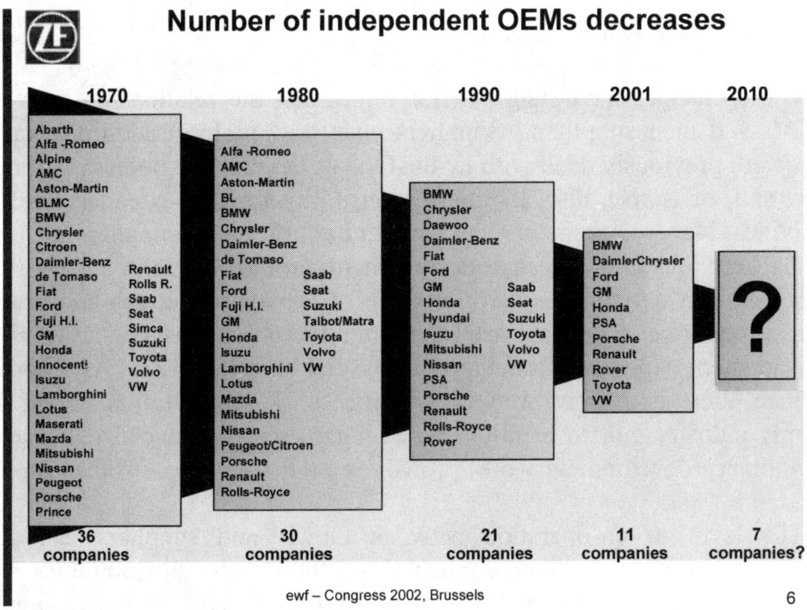

Fig.1: Number of independent OEMs

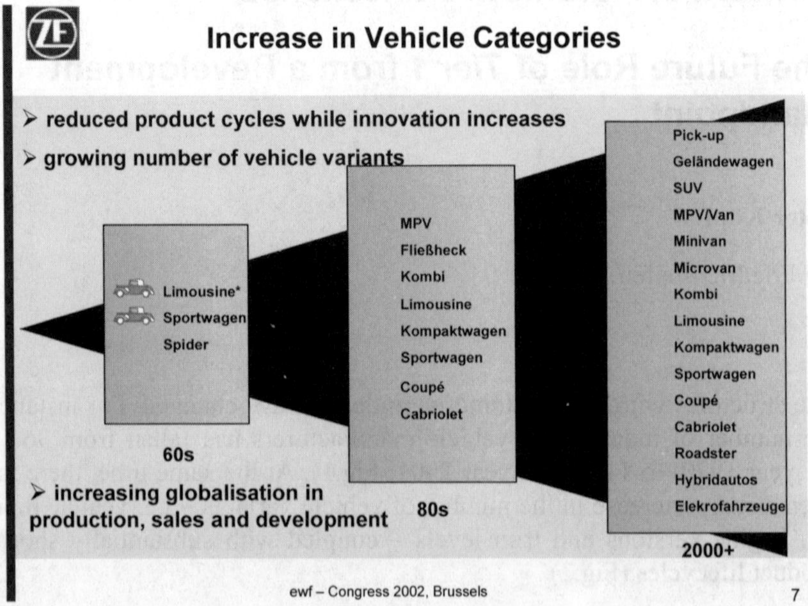

Fig.2: Increase in vehicle categories

These changes have had a major impact on the relationship between OEMs and their suppliers. Suppliers must now perform additional tasks that were previously dealt with by the OEMs, develop the necessary know-how and, of course, also assume responsibility for this extended field of activities. New requirements that have to be fulfilled by suppliers include: global activities in the areas of development, production, sales and service; concentration on core business; creation of flexible structures and procedures; module and system-based thinking; and an increase in the level of development expertise, particularly in the field of electronics. As a consequence, there is a more pronounced process of concentration within the supply industry, not to mention increasingly greater financial risk due to preliminary development work, growth and the take-over of other suppliers.

The level of co-operation between OEMs and suppliers has also changed. Co-operation is now much closer than in the past, involves new IT methods and places new demands on employees. As component suppliers develop into module suppliers and then into system suppliers, so the degree of process integration between OEM and supplier increases. In the case of the system supplier, the goal is close links between all those involved in the development process – with high levels of quality and flexibility – as well as close links within the logistics process, including joint development of work procedures and factories.

Despite all the progress that has been made with respect to IT methods and processes, the problem of the interfaces between different IT systems still remains. There are ideas for solutions involving standardized interfaces. And there are also various IT models that enable co-operation between different companies, some of which prove more advantageous for the OEM, some of which are more beneficial to the supplier. Yet the interface problem is set to become even more serious as e-Engineering continues along the road toward becoming e-Business. Here there remains a wide area in need of standardization and simplification.

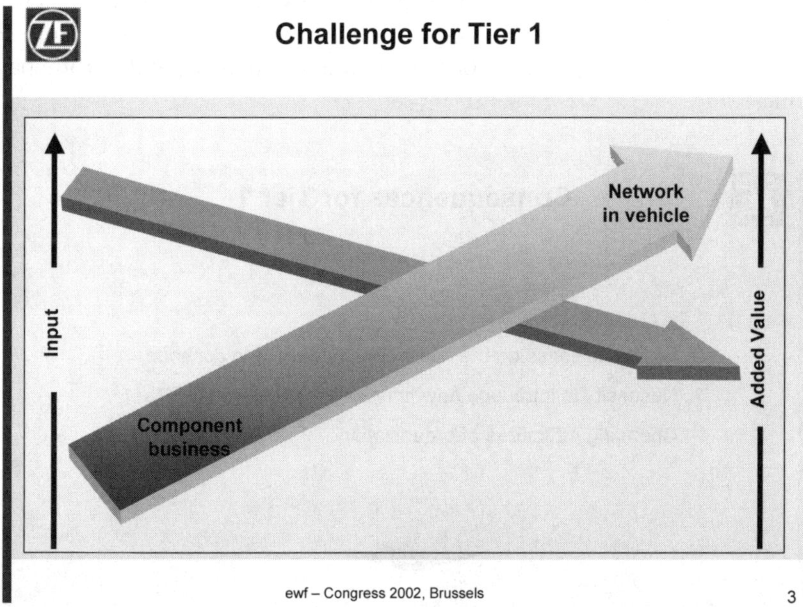

Fig.3: Challenge for Tier 1

ZF Friedrichshafen AG is a company that specializes in driveline and chassis technology. Taking ZF as an example, it is shown how a company can further develop its technological strategy. Existing driveline and chassis hardware components can be transformed into fully-controlled driveline and chassis systems by integrating electronic control processes. Furthermore, the linking of driveline and chassis functionality enables the development of entire dynamic control systems for vehicles.

Electronic systems are an essential part of this development, since they enable the creation of new functions through the networking of individual components, units and systems.

This form of corporate development naturally also has consequences with respect to co-operation with the OEM. Component, module and system business is set to remain the basis of business in the future. This is important as far as added value is concerned. However, it would appear that the development costs factored into the price for basic business will no longer be sufficient to meet the rising expense associated with the development of complex systems. Suppliers therefore face the problem of continually increasing expenditure on the one hand yet ever diminishing added value on the other hand. This situation requires OEMs and suppliers to find a new modus operandi by developing adjusted forms of organization and new methods of settling additional expenses. One possible solution from our point of view involves the supplier putting in place engineering-oriented organizational units. These units would receive orders from the OEM, yet payment for these orders would be kept separate from basic business.

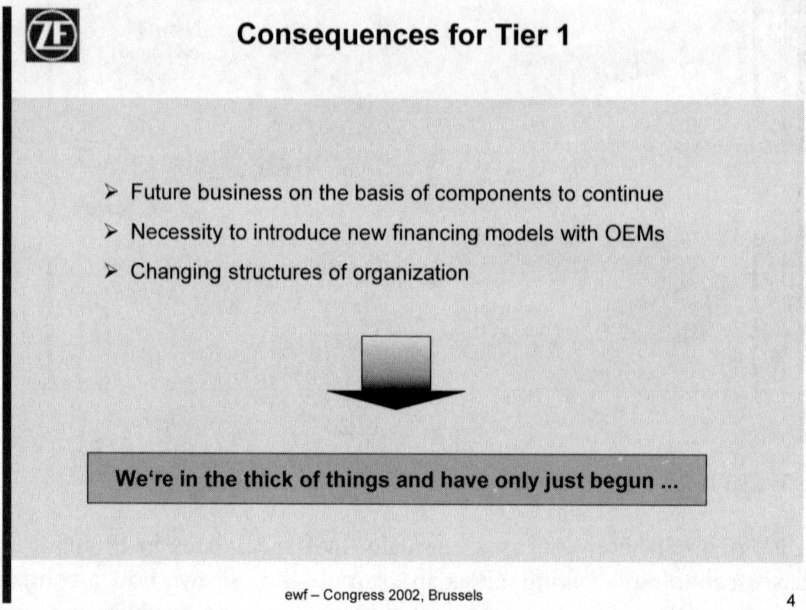

Fig.4: Consequences for Tier 1

eEurope and the European Research Area

EU Policies to Address Major Business and Work Challenges

Erastos Filos

European Commission, Directorate-General Information Society,

eEurope, a political agenda aiming to bring Europe's businesses, citizens and administrations online by 2005, and the European Research Area, targeting creation of a European single market for research, are two major European Union policies that intend to address key challenges for European businesses and workers in the new millennium. The paper shows that R&D efforts under successive European framework programmes for research, development and innovation underpin these policy initiatives and together they offer a coherent approach towards the creation of the largest knowledge-based economy in the world.

1 Introduction

The knowledge economy challenges business in many ways. E-technologies and globalisation are leading to a blurring of organisational boundaries. The creation of value becomes more and more dependent on intangible assets. Uncertain and fast-changing environments require organisational abilities such as flexibility, speed, and adaptability. The networked economy resembles an ecology of organisms, inter-linked and co-evolving, constantly in flux, deeply tangled. It will ultimately change the way businesses relate, both to each other, to the individuals who provide their core competencies, and to their environment.

The Internet responds to the needs of economic actors in the global economy. And it further amplifies globalisation – of the economy, of people, and of ideas. Knowledge, ideas, brainpower are becoming the world's main economic resource. Today, the main added value of a product is in the know-how required to design and market it – and in the services which come along. Intangible goods – content, software, knowledge, etc. – represent an ever-growing share of the economy.

In the past the economy was somehow static, and involved little change. Today, the rapid switchover to the e-economy requires radical changes in attitude and is challenging people's ways and habits: it opens up an era of intense creative thinking, with ideas competing against ideas. The digital age calls for bold minds, an innovative spirit, open-mindedness, and vision. Entrepreneurship at all levels becomes the backbone of all businesses.

2 Challenges for Business and Work

2.1 "Value" Redefined

The Internet and network technologies have led to a new perception of value for individuals and organisations (Zobel & Filos 2002). Firstly, in a network economy the number of possible interactions and options grows exponentially. Also, small efforts can lead to large effects. Mathematically, the sum value of a network increases as the square of the number of members, which means that adding a few members can dramatically increase the value of the network for all its participants, with a self-reinforcing effect. Secondly, in contrast to traditional thinking, where value is related to "scarcity", value in the network economy is seen in the opportunities of relationships. The larger a community becomes, the more value a product acquires as a relationships-enabling means. Also, the more connections there are between members in a net, the more intermediary nodes there can be. Everything in a network is inter-mediating something else. Since a relationship involves at least two members investing in it, value increases exponentially with one's investment. Thirdly, in a digital economy, value is no longer dependent on tangible assets alone. Whereas in the traditional economy value is derived mainly from investment in tangible assets such as plant and machinery, in the knowledge economy "smart" resources such as information, knowledge, brands, relationships, and the capability to innovate become indispensable.

2.2 Trust

A key question of the knowledge economy is: "How can you do business with somebody that you do not see"? (Handy 1995). As business in the knowledge economy relies on technologies and infrastructures that reduce geographical distance, open communication networks and information systems become vulnerable to integrity and security threats that may undermine the very essence of collaboration: trust. Here trust must be established and maintained by security technologies such as cryptography and authentication (e.g. biometrics, electronic signatures, etc.), and by technologies that enhance privacy and help protect and manage intellectual rights, digital assets and identities.

2.3 Relationships

Business success depends on the ability to innovate. Innovation comes from the clash of ideas. The network provides a natural environment for this. Connectivity breeds relationships. The Internet does not only facilitate hyperlinking of documents. People and organisations become "hyperlinked" as well (Levine et al. 2000). The knowledge economy is about the right set of connections between people and organisations, be they customers (*customer relationships management – CRM*) or suppliers (*supply-chain management – SCM*). When relationships are fostered via networks, roles become blurred: the seller becomes the "buyer" of valuable feedback on his product. Mass customisation enables businesses to see customers, suppliers, regulators, and even competitors as stakeholders who can make meaningful and positive contributions (Fig.1).

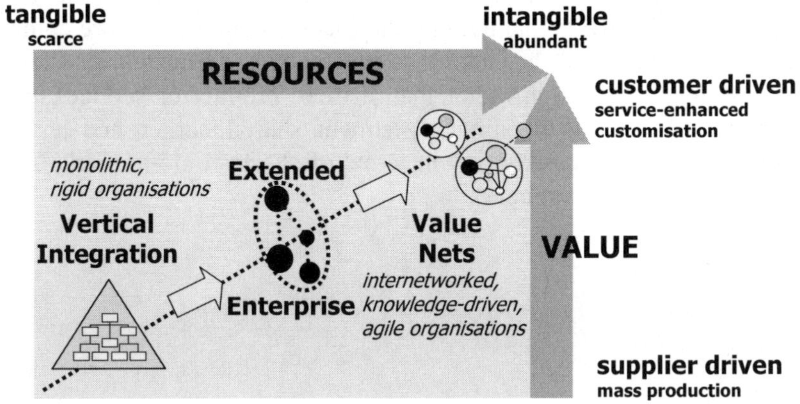

Fig.1: From vertical integration towards the value net (Zobel & Filos 2002)

2.4 Knowledge

Managing knowledge becomes an ever growing challenge for businesses and individuals. The increasing complexity of products, processes and networking relationships will require management skills based on the ability to leverage organisational knowledge, including its tacit dimension. The knowledge economy empowers individuals and opens new possibilities in particular for the smart SME.

2.5 Internetworking

For businesses, large and small, collaborative partnerships have become central to competitive success in fast changing global markets. Since many of the skills and resources essential to an organisation's competence lie outside its boundaries, and outside management's direct control, partnerships are not an option anymore but a necessity. Organisations today have to be "smart" in their ability to conceive, shape and sustain a wide variety of collaborative partnerships. Hence the challenge: the "capacity to collaborate" becomes a core competence of organisations. Collaborative partnerships are held together because of the added value they offer. There is a variety of strategic goals that organisations may pursue by entering into co-operation with others, such as, *resource optimisation* (sharing investment with regard to infrastructure, R&D, market knowledge, etc. and the sharing of risks, while maintaining the focus on one's own core competencies), *synergy creation* by linking complementary competencies (i.e. to offer customers a solution rather than a mere product or service), *achieve critical mass* in terms of capital investment, shared markets and customers; *achieve increased benefits* (i.e. in terms of shorter time-to-market, higher quality, with less investment).

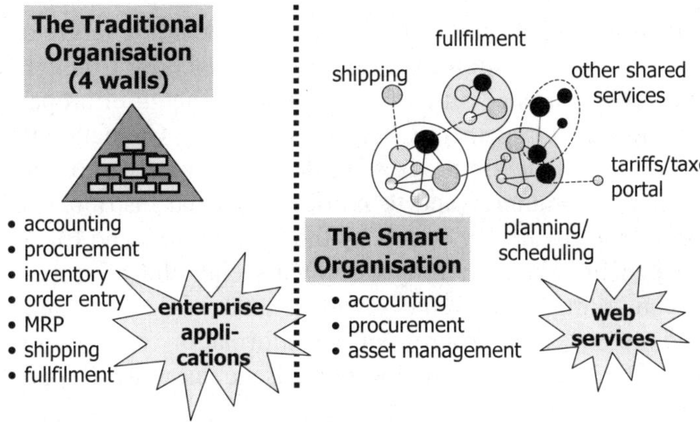

Fig.2: Traditional versus internetworked business (Zobel & Filos 2002)

As the economy is "going digital" businesses are starting to realise what implications this will have on business processes and on the organisation as a whole. The transformation involves much more than setting up a digital infrastructure and requires more than the capability to enter into a virtual collaboration with other partners (Filos & Banahan 2001). Whereas in the past the aim for organisations was to integrate the supply-chain as tightly as possible, the focus is shifting from the vertically integrated towards the *internet*worked organisation. One of its characteristics is its focus on the core business while outsourcing other activities via the Internet and e-business exchanges to partners that may have the expertise to perform specific tasks better and more cost-effectively (Fig.2).

3 The Lisbon Strategy

In 2000, the Member States of the European Union adopted a strategy (Lisbon Council 2000) with the objective for Europe "to become the most competitive and dynamic knowledge-based economy in the world, capable of sustained economic growth, with more and better jobs, and greater social cohesion" (Fig.3).

3.1 The eEurope Initiative

At the Lisbon Summit in March 2000, heads of state and government committed themselves to a number of measures, including target dates, to bring eEurope forward (Action Plan 2000). The Göteborg Council, a year later, adopted the eEurope+ initiative that complements eEurope by extending activities to Enlargement Countries (Göteborg Council 2001). The initiative set out to accelerate progress in the legal framework on telecommunications and e-commerce and to reorient financial instruments. Common legislative and regulatory efforts by the Commission and EU Member States were benchmarked on progress. Overseeing the eEurope Action Plan, a benchmarking Programme - with a total of 23 key indicators - monitors and measures the progress of the Information Society in Europe. The eEurope 2002 Benchmarking Report (Benchmarking Report 2002) presented first results and achievements, e.g.:

- A new telecommunications package, including a regulation on "unbundling the local loop", aiming to generate improvements for consumers and industry in Europe, were rapidly adopted;
- A light, but effective, legal framework on e-commerce was set up;
- Internet penetration increased from 18% in March 2000 to 38% in December 2001;
- Nearly 90% of schools are now connected to the Internet and over 90% of businesses have Internet access; governments are moving online;
- Europe now has the fastest research network in the world.

Some elements of the eEurope Action Plan (Action Plan 2000) are:

3.1.1 Dependability of information infrastructures

The Information Society is increasingly relying on infrastructures for life-critical and business-critical functions. The wording «information infrastructure» is used to collectively indicate the ensemble of media, network, and communication infrastructures up to the application level. New and more sophisticated instances of attacks (e.g. virus, denial of service) are growing. The economical damages caused by disruptions in the functioning of the Internet are increasing. In response to this scenario, a series of specific targets on secure networks have been set in the eEurope 2002 Action Plan. One of these is to improve co-operation amongst national Computer Emergency Response Teams (CERTs).

3.1.2 Secure networks and smart cards

eEurope proposes to improve the security of on-line transactions by supporting the development of certification services and Internet security solutions and by encouraging the development of common specifications for smart cards. In January 2002 the European Council adopted a resolution on a common European approach and specific measures for «Network and Information Security» (Security 2002), to which the Information Society Technologies programme (IST Key Action II 2002) provides the technological support. Smart cards are private, personal and secure objects possessed and directly controlled by citizens. Accordingly, they are perceived as trusted tokens holding their specific data. Smart cards operating as «trust enabling» tokens can act as key elements in providing open and friendly access to applications and services of the Information Society. As part of the overall eEurope initiative a Smart Card Charter (Smart Cards 2002) and common requirements were devised. Both were endorsed by a large constituency of 450 organisations involving major card issuers, the chip card industry and user communities. They also identified issues that need to be resolved before smart cards can fulfil their potential to fully support the aspirations of citizens with respect to technology (Smart Cards 2002). The Smart Card Charter addresses the needs of citizens and the business community in terms of business cases, multi-functionality and interoperability of systems and infrastructure, and the provision of trust in all aspects of service delivery.

3.1.3 Promoting privacy-enhancing technologies and best practices

Besides the technical solutions offered by projects launched in the area of privacy enhancing technologies, policy activities involve the development and animation of the e-Forum on Privacy in Information Society (Privacy 2002). It is aimed to become a portal for awareness activities and exchange of experiences and best practices with respect to improving security, ensuring privacy, preventing and protecting from attacks.

3.1.4 Work in a knowledge-based economy

The European strategy for growth and employment, adopted at the Lisbon Summit in March 2000 sets the challenging goal of increasing participation in employment to near 70% by 2010 (Lisbon Council 2000). This requires action to improve the employment prospects of groups with low employment rates, especially women and older workers. Work can be made more attractive and accessible through flexible work arrangements such as e-work. Particular efforts should be made to attract women to the knowledge economy and IT professions where they are massively under-represented and where they constitute a largely untapped resource in most countries. This must be accompanied by the modernisation of work organisation. Social partners are encouraged to support agreements on flexible working to the advantage of both employers and employees.

3.1.5 GoDigital for SMEs

To accelerate the take up of e-business in Europe, the eEurope Action Plan foresees to "encourage SMEs to 'Go-Digital' through co-ordinated networking activities for the exchange of knowledge on best practices, e-commerce readiness, and benchmarking". GoDigital is an EU initiative supporting SMEs doing business online. Specific objectives of this initiative (GoDigital 2001) are:

- To identify the main obstacles SMEs face as they engage in e-business;
- To propose specific actions to help SMEs «go digital», in particular by building on existing policies and initiatives;
- To ensure consistency among the various policies and initiatives launched for supporting SMEs going digital at the European, national, regional and local levels;
- To learn from practical experience and to benchmark various strategies to help SMEs going digital.

3.2 eEurope 2005

Building on the achievements of the *e*Europe 2002 discussed above, EU leaders decided at the European Council in Sevilla in March 2002 to launch *e*Europe 2005. The objective of the new action plan is to provide a favourable environment for private investment and for the creation of new jobs, to boost productivity, to modernise public services, and to give everyone the opportunity to participate in the global information society. *e*Europe 2005 therefore aims to stimulate secure services, applications and content based on a widely available broadband infrastructure. The action plan (Action Plan 2002) is based on two groups of actions that reinforce each other. On the one hand, it aims to stimulate services, applications and content, covering both online public services and e-business; on the other hand it addresses the underlying broadband infrastructure and security matters. The action plan comprises four separate but interlinked tools. Firstly, policy measures to review and adapt legislation at national and European level; to ensure legislation does not unnecessarily hamper new services; to strengthen competition and interoperability; to improve access to a variety of networks; and to demonstrate political leadership. *e*Europe 2005 identifies those areas where public policy can provide an added value and therefore focuses on a limited set of actions in priority areas. Some key targets are:

- Connecting public administrations, schools, health care to broadband;
- Interactive public services, accessible for all, and offered on multiple platforms;
- Provide online health services;
- Removal of obstacles to the deployment of broadband networks;
- Creation of a dynamic e-business environment;
- Creation of a Cyber Security Task Force.

Secondly, *e*Europe 2005 will facilitate the exchange of experience, of good practices and demonstration projects, but also of sharing the lessons from failures. Projects will be launched to accelerate the roll out of leading edge applications and infrastructure. Thirdly, policy measures will be monitored and better focused by benchmarking of the progress made in achieving the objectives and of the policies in support of the objectives. Fourthly, an overall co-ordination of existing policies will bring out synergies between proposed actions. A steering group will provide a better overview of policy developments and ensure a good information exchange between national and European policy makers and the private sector. This steering group would also make an early participation of candidate countries possible.

3.3 Sustainable Development

There is a growing public and political awareness that sustainable growth can be realised by using material and natural resources more efficiently. This can be achieved by improving processes, products, product to service conversion, and structural change. *Process re-engineering for e-business* has been given a major stimulus by the emergence of multimedia information infrastructures. *Product improvement* has come through use of new materials, and through the increase of information content in products. Many products can be substantially de-materialised. On the one hand, with advanced communications, *products become services*: a newspaper becomes an on-line news service; an instruction manual becomes an interactive technical advice service. On the other hand, *structural changes* in the way markets and transport infrastructures are organised and used, and in the way people work and live, promise great benefits.

A set of case studies of how the transition to a knowledge economy can contribute to resource-use efficiency and sustainable development were published in May 2000 (Forseback & Johnston 2000), together with the proceedings of the Conference "Towards a Sustainable Information Society" held in February 2000 (Johnston et al. 2000).

Fig.3: Key policy areas in the framework of the Lisbon Strategy (Zobel & Filos 2002)

3.4 Towards the European Research Area

In January 2000 the European Commission proposed to create the "European Research Area" (ERA 2000) to contribute to better framework conditions for research in Europe by combining efforts undertaken on European level with those in Member States. This was confirmed at the Lisbon Council of 2000. Two years later, at the Barcelona European Council, which reviewed progress towards the Lisbon goals, EU heads of state and government agreed that R&D investment in the EU must be increased with the aim of approaching 3 % of GDP by 2010, up from 1.9 % in 2000 (ERA 2002). They also called for an increase of the level of business funding, which should rise from its current level of 56 % to two-thirds of total R&D investment, a proportion already achieved in the US and in some European countries. This twin objective is ambitious but realistic: today, several European countries are close to or beyond these levels. The 2002 Broad Economic Policy Guidelines of the Member States and the Community acknowledge the importance of this goal and recommend to improve incentives for firms to invest in R&D while preserving sound fiscal policies. The R&D investment objectives set at Barcelona arise from the recognition that strengthening our R&D and innovation systems is essential in realising the Lisbon strategic goal. Its achievement is put at risk by the large and growing gap in R&D investment between the EU and the US. This gap reached more than € 120 billion in 2000, 80 % of which was due to lower R&D investment by business in Europe. The role of R&D as a driving force for a competitive and dynamic knowledge-based economy is linked to the economy's capacity to turn new knowledge into technological innovation. Although many enterprises recognise the increased importance of investing in R&D, they will do so only to the extent that they can exploit results effectively and expect sufficient returns to balance the risk inherent in such investment. The proposal identifies the wide range of policy areas which must be mobilised in a coherent manner. It sets out in each area the main objectives to be pursued either by intensifying actions already underway in the context of the Lisbon strategy and the European Knowledge Area or by undertaking new initiatives. Even where action has been launched and agreed at European level, more needs to be done to ensure that it is delivering results at national and local levels. At the same time, it is recognised that the diversity of situations in Member States and Candidate Countries must allow for a differentiated policy response. More attractive framework conditions are necessary. Among the most important in this respect are a sufficient supply of highly qualified human resources, a strong public research base, a dynamic entrepreneurship culture, adequate systems of protecting and managing intellectual property rights, a competitive environment with research and innovation-friendly regulations and competition rules, supportive financial markets, macro-economic stability and favourable fiscal conditions. There is also a case for a more effective and focused use of public financial incentives to private R&D and technology-based innovation, within the context of state aid rules, which imply that efforts to enhance public support for R&D must come to a large extent through restructuring of public expenditure. In this respect, public authorities have a range of financing instruments at their disposal, in particular direct support measures, fiscal incentives, guarantee schemes and public support for risk capital. A mix of these instruments is required, as no single instrument is able to provide the full range of incentives. Lastly, the place of R&D in the overall business strategy of companies as well as the effectiveness and efficiency of their R&D activities are important factors to consider.

4 R&D Activities to Address Business and Work Challenges

4.1 Activities of IST Key Action II

Since the beginning of the IST Programme in 1999, some 450 projects with a funding budget of half a billion euro have been launched in the area of New Methods of Work and Electronic Commerce (Zobel & Filos 2002, IST Key Action II 2002). They address the following areas:

4.1.1 Technologies and infrastructures for trust and security

They embrace a wide range of techniques like cryptography, authentication mechanisms and infrastructures (including electronic signatures), privacy enhancing technologies, tamper-proof hardware and components, and support the development of emerging business models. They also empower users to effectively manage their personal rights, they help to prevent and fight abuses. Once building blocks are properly developed, integrating them in robust infrastructures, and testing a smooth operation is a long undertaking. These issues range from the scalability of the solutions to the legal impact of the implementations. *Secure financing and transactions* projects address mainly the security of financial electronic transactions and hardware securing technology, while leaving the business process mainly uncovered. In this area the smart card is also a vehicle of choice for secure financial deployment. *Digital rights management* concentrates on electronic transfer and trade of digital content to which rights are associated. In the knowledge economy content is sold in digital form and the complete sales operation (advertisement, catalogues, negotiation, contracting, payment and delivery) is carried out electronically. Research on *smart cards* goes far beyond the cards themselves: it brings together technology companies and payment solution providers in order to specify a European standard for secure smart card readers (FINREAD 2002). *Dependability* activities aim at fostering a dependability-aware culture in developing, managing and operating the whole life-cycle of technical and business components, systems, networks and infrastructures (DEPPY Forum 2002).

4.1.2 Business organisation and supply-chain management

Smart organisations, i.e. knowledge driven, adaptive and learning as well as agile in their ability to create and exploit the opportunities of the *internet*worked economy are key signposts of the shift from the industrial to the digital era. These organisations are virtual in concept, highly flexible, dynamic, and capable of leveraging the power of network technologies to meet customer demands for high added-value products and services in a global market. R&D projects in this area explore and validate technologies and architectures to develop platforms for interoperability of organisational applications, tools to manage knowledge in organisations, mobile applications, innovative models for process distribution and re-engineering models for networked organisations (Filos 2002). However, technology is only an enabler. The challenge is in using technology to its full potential. Projects selected in 1999 concern the improvement of functional integration within and between organisations, both in volatile virtual arrangements and in more permanent supply-chain configurations. Many projects were launched in order to provide early benefits to SMEs, i.e. a sensible balance between innovation and usability. The R&D focus shifted later to process distribution and interoperability of heterogeneous enterprise software, incorporating intelligence into business processes, and developing applications for wireless and mobile environments. *e-Business* projects deal with the dynamic creation of highly customised products and services in response to changing market demands and electronic mediation to identify and select value constellation partners (e.g. via automated negotiation, auctioning and agent-based contracting). R&D includes issues such as life cycle management of highly customised products and services across dynamic value constellations (i.e. from product conception all the way to product dismantling and resource recovery), extended products (which combine tangible and intangible elements), customer relationships management and online solutions for alternative dispute resolution. *Knowledge management* projects address technologies in the context of organisational knowledge management, teamwork and business collaboration, thus aiming to enhance effectiveness and co-operation within and between organisations (KM Forum 2002).

4.1.3 New methods of work

The move from telework towards *e-work* represents a new phase for the European workforce. Better working environments, better working conditions, and a better reconciliation of work and personal life are the objectives. *Mobile communications* are increasingly dominating the new generation of e-commerce and e-work systems. Despite all their technology and market issues, the emerging GPRS and UMTS platforms will be of major importance in the near future. However, research is still required to fine-tune the technology solutions, and to better comprehend the deployment difficulties and opportunities (Mobilforum 2002). In workplace design the focus is more tightly on multidisciplinary R&D for *sustainable workplace design*, including the architectural and resource-efficiency issues. This work will continue to address multidisciplinary development for workplace and office design. The inclusion of all people who might be marginalized by the digital divide is becoming more prominent. In addition, the link with the new ways of working is extremely strong. These activities enhance sustainability through providing the platforms and tools needed in future dynamic, satisfying, safe, secure, mobile and distributed workplaces. They may lead to improved resource use efficiency in the workplace and built environment, e.g. through replacing physical prototyping by virtual prototyping or establishing highly utilised offices in the vicinity where people live (Filos 2002).

4.1.4 Socio-economic research

Successful development requires a strong interplay between the technical, economic, social and legal issues. The core of *socio-economic research* activity, however, is implemented through 50 projects contracted so far in IST Key Action II (IST SER 2002). The interdisciplinary character of this field of work is particularly challenging in terms of setting and achieving focused objectives in a fast moving and highly dynamic environment. Projects cluster around three distinct fields, such as measuring the new economy, identifying shaping factors of the emerging new ways of work and business and contributing to supporting EU policies.

4.2 Two Project Examples

The European R&D programmes have played a significant role in supporting the development of key technologies and applications that helped create the Information Society. Key to the emergence of the knowledge economy are continuous R&D efforts in virtual business technologies, concurrent engineering, computer-supported collaborative work (CSCW) and product and process data technologies. For example, concurrent engineering has been a priority within the European Commission's R&D programmes since the early 90's. More than 70 industry-led projects receiving more than 300 million euro of support (shared cost funding, 50% industrial contribution) have been set up, some of the most relevant being ENHANCE and AIT mentioned below (Filos 2001).

4.2.1 Enhanced Aeronautical Concurrent Engineering (ENHANCE)

The ENHANCE industrial consortium groups together the main European civilian aeronautical companies EADS (formerly Aerospatiale-Matra, CASA, DaimlerChrysler Aerospace), British Aerospace, Dassault Aviation and Alenia, engine manufacturers such as SNECMA and Rolls Royce, plus equipment manufacturers and airlines, SMEs and European research centres (ENHANCE 2002). The objectives were:

- To draw up common methods and standards, applicable from the initial design phases throughout the entire life of an aircraft, as part of a general concurrent engineering methodology;
- To establish a common definition of the aeronautical development framework (with all its aspects related to processes, methodologies and organisation);
- To define baselines: terminology, item identification, models and general data exchange specifications and a common methodological support;
- To set up pilot projects designed to demonstrate the credibility of the methods and means adopted to consequently validate their industrial effect, through the operational application of concurrent engineering techniques;
- To investigate the operation of simple consultancy structures to assist the project regarding the analysis of human factors (resistance to change, evolution in organisation, training, etc.) and information technology.

The project has set up successfully most of the pilot applications and is working towards its exploitation phase.

4.2.2 The Advanced Information Technology (AIT) initiative

The initiative was launched in 1993 by the major European automotive companies with support from the European Commission. It was organised in two distinct phases: the pilot phase, with a duration of 24 months, consisted of three steps, namely identification of end user industrial needs, consensus building, and specification of requirements. The main phase focused on performing the necessary R&D work and intended to provide enabling IT tools and techniques for the industrial requirements, as specified in the pilot phase. Almost 100 organisations participated in the initiative with a total budget in the range of 130 million euro (Garas & Naccari 2001). The main objectives of AIT were:

- To provide industrial leadership to facilitate the co-operation between IT users, IT vendors, research organisations, academia and standardisation bodies;
- To identify and define technical domains and user requirements for innovative IT products;
- To propose appropriate R&D projects relevant to the needs of the involved industries;
- To promote international standards and to contribute to their development;
- To jointly develop and utilise effective organisational, technical and operational guidelines for co-operation.

5 R&D Activities Foreseen under the 6th Framework Programme

The 6th Framework Programme for Research (FP6 2002), which sets out the priorities for research on European level in the years 2002-2006, aims at bundling and focusing these research efforts towards the objectives of the European Research Area. The goal of this programme is to promote world-class research in key priority areas of exceptional interest and added value to Europe through activities such as,

- Scientific and technological research, development and demonstration;
- Co-operation with researchers and research teams from third countries;

- Dissemination and exploitation;
- Human resource development, including the promotion of training of researchers;
- Development of research facilities and infrastructures for research;
- Promotion of better links between science and society.

The new programme's envisaged eight research areas and topics are: genomics and biotechnology for health, information society technologies, nanotechnologies, intelligent materials and new production processes, aeronautics and space, food safety and health risks, technologies for sustainable development and transport, citizens and governance in the knowledge-based society, and issues regarding the new and emerging scientific and technological developments of relevance to Europe. Issues relevant to business and work will be found under the following 6th Framework Programme:

- *In priority area 2 - Information society technologies:* The relevant part is "Networked businesses & governments" and "Products and services engineering 2010" (FP6 Workprogramme 2002);
- *In priority area 3* – Nanotechnologies and nano-sciences, knowledge-based functional materials, new production processes and devices;
- *In priority area 4 - Aeronautics and space:* The relevant part here is the "aeronautics" thematic area, dealing with "strengthening competitiveness by reducing development costs, aircraft direct operating costs and improving passenger comfort"

6 Conclusions

Policies, such as eEurope and the European Research Area, have drawn on the vision for Europe becoming "the largest knowledge-based economy in the world" by the year 2010, which was formulated in Lisbon. The European R&D Programmes continue to play a major role in supporting the development of technologies and applications that help bring about and shape the Information Society and the knowledge economy. Besides notable R&D achievements, important progress has also been made in the area of standardisation and consensus building between the research, business communities and societal groups in Europe. It is recognised that technology alone will not be able to sustain the creative potential of the knowledge economy and that both are needed to bundle the efforts into a coherent strategy in addressing the business and work challenges of the new millennium.

7 Acknowledgements

The views expressed in this article are those of the author and do not necessarily reflect the official European Commission's view on the subject.

8 References

Action Plan (2000), 'eEurope 2002, An Information Society for All, Action Plan', Brussels, 14 June, available under: http://europa.eu.int/ information_society/eeurope/action_plan/pdf/actionplan_en.pdf

Action Plan (2002), 'eEurope 2005: An Information Society for All, An Action Plan to be presented in View of the Sevilla European Council', COM (2002) 263 final, 28 May, http://europa.eu.int/information_society/eeurope/ news_library/documents/eeurope2005/eeurope2005_en.pdf

Benchmarking Report (2002), 'eEurope 2002 Benchmarking Report', COM (2002) 62 Final, Brussels, 5 February, available under http://europa.eu.int/information_society/eeurope/news_library/ new_documents/benchmarking/benchmarkin_en.pdf

DEPPY Forum (2002), 'European Dependability Initiative', 30 November, http://deppy.jrc.it

Enhance (2002), 'Enhance. Bulding Europe's aeronautical future', 30 November, http://www.enhanceproject.com/

ERA (2000), 'Towards a European Research Area', COM (2000) 6 final, - 18 January, available electronically under, http://europa.eu.int/comm/research/ era/pdf/com2000-6-en.pdf

ERA (2002), 'More research for Europe. Towards 3% of GDP', COM (2002) 499 final, - 11 September, available electronically under, ftp://ftp.cordis.lu/pub/ documents_r5/natdir0000023/s_1869005_20020913_174005_GERC021869 en.pdf

Filos, E. (2001), 'The Future of Concurrency', CE2001, Anaheim, California, USA, 28 July - 1 August, in: Roy R. and Prasad B. (eds), Advances in Concurrent Engineering, Proceedings of the 8th ISPE International Conference on Concurrent Engineering: Research and Applications, ISBN 0-9710461-0-7

Filos, E. and Banahan, E. (2001), 'Towards the Smart Organization. An Emerging Organizational Paradigm and the Contribution of the European RTD Programmes', Journal of Intelligent Manufacturing, 12, 2, pp 101-119

Filos, E. (2002), 'European collaborative R&D projects related to the "Smart Organisation". A first evaluation of activities and implications for construction', 4th European Conference on Product and Process Modelling in Building and

Construction (ECPPM), 9-11 September, Portoroz, Slovenia, in: Turk & Scherer (eds) (2002), 'eWork and eBusiness in Architecture, Engineering and Construction', Sweets & Zeitlinger, Lisse, ISBN 90 5809 507 X, pp. 27-32

FINREAD (2002), 'Europe Unites to Provide e-Business Security', 30 November, http://www.finread.com

Forseback, L. and Johnston, P. (2000), 'Case Studies of the Information Society and Sustainable Development', May, Brussels, European Commission, Directorate-General Information Society, available electronically under: ftp://ftp.cordis.lu/pub/ist/docs/ka2/casesfinal.pdf

FP6 (2002), 'The sixth framework programme of the European Community for research, technological development and demonstration activities, contributing to the creation of the European Research Area and to innovation (2002 to 2006)', Official Journal of the European Communities, L232, 29 August, pp. 1-33, available electronically under, ftp://ftp.cordis.lu/pub/documents_r5/natdir0000029/s_1831005_20021107_150652_6FPL021654en.pdf

FP6 Workprogramme (2002), to be announced in December 2002, under: http://www.cordis.lu/FP6

Garas, F.K., Naccari, F. N. (2001), 'Assessment of the AIT Initiative', Report to the European Commission, Brussels, available under: http://www.cordis.lu/ist/ka2/rptspolicyconf.htm

GoDigital (2001), 'Helping SMEs to "Go Digital"', COM (2001) 136 final, 13 March, http://europa.eu.int/information_society/topics/ebusiness/godigital/docs/background/en.pdf

GoDigital (2002), 'Getting Europe On-line and Doing e-Business', 30 November, http://europa.eu.int/godigital/

Göteborg Council (2001), 'Presidency Conclusions. Göteborg European Council', 15-16 June, available under, http://europa.eu.int/council/off/conclu/

Handy, C. (1995), 'Trust and the Virtual Organization', Harvard Business Review, May-June, pp. 25-50

IST Key Action II (2002), 'Key Action 2, New Methods of Work & Electronic Commerce, Projects', 30 November, http://www.cordis.lu/ist/ka2/projects.htm

IST SER (2002), 'Key Action 2, New Methods of Work & Electronic Commerce, Socio-economic Research (SER) Activities', 30 November, http://www.cordis.lu/ist/ka2/ser.htm

Johnston, P., Bohlin, E. and Clements, B. (2000), 'Towards a Sustainable Information Society', Report of Conference of 21-22 February, June, Brussels, European Commission, Directorate-General Information Society and Joint Research Centre - IPTS, available electronically under, http://www.jrc.es/pages/projects/AsisProceedingsFinal.pdf

Kelly, K. (1999), New Rules for the New Economy, Forth Estate, London

KM Forum (2002), 'KnowledgeBoard: The European Knowledge Management (KM) Community - Home', 30 November, http://www.knowledgeboard.com/

Levine, R., Locke, Chr, Searls, D. Weinberger, D. (2000), The Cluetrain Manifesto. The End of Business as Usual, Perseus, Cambridge, Mass.

Lisbon Council (2000), 'Presidency Conclusions. Lisbon European Council', 23-24 March, available under, http://europa.eu.int/council/off/conclu/

Mobiforum (2002), 'm-Commerce Portal', 30 November, http://www.mobiforum.org/

Privacy (2002), 'ePrivacy Forum', 30 November, http://eprivacy.jrc.it

Security (2002), 'Council Resolution on a Common Approach and Specific Actions in the Area of Network and Information Security', 2002/C 43/02, 28 January, Official Journal of the European Communities, C43, 16 February, pp. 2-4, available electronically under, http://europa.eu.int/information_society/eeurope/action_plan/safe/netsecres_en.pdf

Smart Cards (2002), 'eEurope Smart Cards', 30 November, http://eeurope-smartcards.org

Zobel, R. and Filos, E. (2002), 'Work and Business in the e-Economy. Technology & Policy Issues', 3rd European e-Business and e-Work Conference, Prague, Czech Republic, 16-18 October, in: B. Stanford-Smith et al. (eds) (2002): Challenges and Achievements in E-business and E-work, IOS Press, ISBN 1 58603 284 4, pp. 52-66

Networks of Excellence: Relevance, Feasibility, Sustainability

Peter Sachsenmeier

Member of the Executive Board, IMAG Information Management AG, Schaffhausen, Switzerland; Vice President ewf Europe

An Exciting Idea, Three Fundamental Tests

The mention of the expression "Network of Excellence" (NoE) never fails to make people curious, arouses their interest, it seems such an exciting idea.

NoEs belong to an arena of thought which is currently thriving. Competitive strategy dominated the debate in the 1980s, cooperative strategies became fashionable in the 1990s. We are now at a point where we want to blend the two, into a greater whole.

Putting this idea into practice, however, appears to be more difficult. Expectations vary, for one thing, and if we cannot manage these appropriately, we shall fail in our attempt at establishing successful NoEs. Such expectations include the rather romantic notion that Networks of Excellence are another silver bullet with which to cure many ills, and turn to everybody's advantage the tectonic changes underway in many global industries.

In a book chapter elsewhere on the failure of the dot.coms (Sachsenmeier, 2002), I stipulated three simple yet fundamental questions to test any grand new idea in order to avoid being blinded by its glamour, and to investigate its true potential:

1. Is it relevant for those who'll have to pay for it?
2. Is it feasible? Are we going to be able to manage the processes, the necessary changes, and the technologies?
3. Is it sustainable? Can we ensure it will remain, survive, flourish?

The bulk of this chapter is devoted to these fundamental questions; the relevance, feasibility, sustainability of NoEs. Also, I shall provide some ideas, models, guidelines and checklists in order to illustrate the possible workings of NoEs.

I am keenly aware that collaboration is not easy, and the creation of value in an NoE is a very difficult task. Successful collaboration boils down to trust. And trust has to be built up over time. Good methods for achieving it are to start small, "singing" the benefits to all those involved internally, using dialogue in old-fashioned meetings, going for the win-win, settling proprietary information issues, and just doing it. Having a history of successes together goes a long way to maintaining trust!

Definitions and Explanations

People want to know what it all means. Let us start with an exegesis of the term "Network of Excellence".

ewf's selection of the term "Network of Excellence" owes much to the European Union's 6th Framework programme (EU FP6). However, over time, the term became a description for a field of activities in which our association, ewf europe, wanted to be active and pursue its own set of programmes.

In EU parlance, a Network of Excellence (NoE) is "an instrument for strengthening excellence by tackling the fragmentation of European research, where the main deliverable is a durable structuring and shaping of the way that research is carried out on the topic of the network."

The aims and objectives of such EU defined NoEs are to strengthen scientific and technological excellence on a particular research topic. The fragmentation of European research is supposed to be overcome by networking together the critical mass of resources, and by networking the expertise needed to provide European leadership. NoEs are expected to have a mandate to spread excellence beyond the boundaries of their partnership, and may include a training component.

The EU stipulates that there be a minimum of three partners from three different countries, and that the duration of funding for an NoE be a maximum of 5-7 years. An overall management framework is expected to be in place.

ewf is going to take the EU's experience with its programme into account and we shall learn from it. We expect to move beyond the EUs prescriptions, into the uncharted waters of NoEs devoted to specific issues in our key industries, automotive, aerospace, and many others.

The way we look at excellence determines the expectations we have of NoEs. The term excellence has been much used, and confused. In 1982, Tom Peters and Robert Waterman wrote their best-selling book, In Search of Excellence. In this, they identified 43 excellent companies based on six financial yardsticks (compound asset growth, compound equity growth, ratio of market value to book value, return on capital, return on equity and return on sales). Five years hence, more than two thirds of those so-called "excellent" companies floundered. In business and management studies, many were shocked and continue to wonder what the elusive concept of excellence is all about.

Excellence will remain confusing until we realize it is not a one-off, isolated item. While dictionaries define excellence in terms of going beyond a limit or standard, and of being superior, we in business and industry have long realized that excellence is a habit, a continuous improvement process, not a goal post.

Excellence is about the continuous striving for being the best, being the first, and about being different and outstanding in the workplace and in the marketplace every day, in every way possible. The aim is to win over the customers, and to make them perceive and judge everything we produce, sell or do as being excellent.

Since excellence is a matter of perception, it is also an interactive process.

Questions remain:

- How far can we go in (pre-) defining the NoEs?
- Can structures that have hitherto been perceived as – often hierarchical – customer-supplier relationships aspire to become NoEs, both with proper networking and excellence characteristics?
- What must we do, how can we support such networks in order to become recognized by their various publics as truly excellent?
- From our look at NoEs as part of our programme, can we truly identify methods and procedures which will help others to become NoEs more quickly, more productively, more cheaply?

We hope to be able to have answers to these legitimate questions once we have a few successful NoE projects under our belt.

A Holistic View of the Life Cycles

A lot of effort currently goes into *forming* NoEs, and people think hard about how to *run* them. We must not forget that networks typically have a lifecycle of *three* major phases (Fig.1). Two of these, the beginning ("engage") and the end ("dissolve") tend to be underestimated as to the resources and effort required.

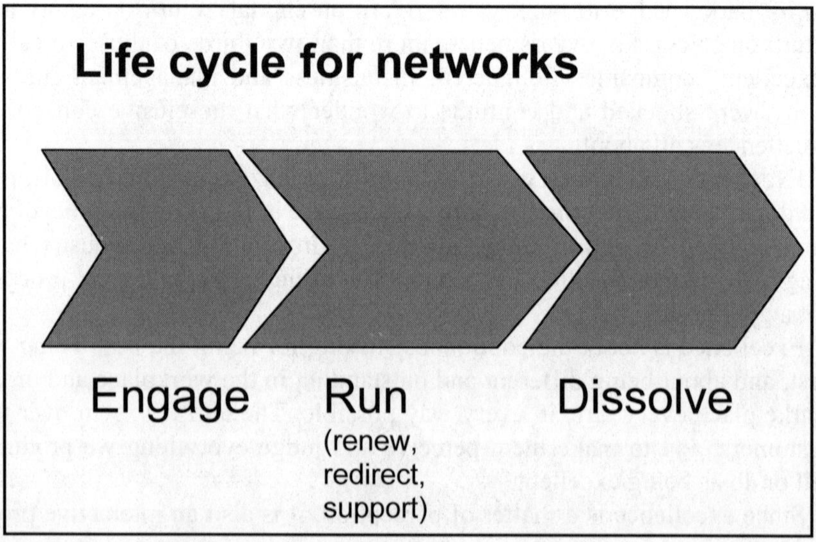

Fig.1: Life cycles for networks

In the rush to establish such networks, the long identification and engagement phase, the very substantial and involved effort needed to run them and the final, huge effort to dissolve them to everybody's satisfaction are all typically underestimated. In some instances, for example in Canada's Networks of Centers of Excellence, even the engagement phase was found to be one year or longer. We - as ewf - have also expended considerable "brainware" on initiating and running such NoEs but as yet little or none on renewing, redirecting and dissolving them

As ewf, we shall help to make the entire life cycles of "our" NoEs a success, through a pertinent programme management (of which more later).

Relevance

Rapid economic and technological change, shifting patterns of productivity, geographical redistribution of economic growth, increasing competitive pressures coupled with a scarcity of essential resources and know how, global interdependence, the blurring of boundaries between distinct legal entities, all of these lead to collaboration.

Collaboration is a deep-seated human response to turbulence which individuals or individual organisations are unable to manage.

However, in all business situations, there are key drivers which determine the exact goals and scope of collaborative ventures. Process, capability, value proposition and channel have been proposed as the four key drivers which lead to collaborative structures. For business processes, supply chain networks emerge (process). Missing or core capabilities demand a capability-based collaboration (capability). The development of new value propositions requires proposition-based collaboration. Efforts to control channel to market need yet another type of collaboration (channel). Goals and scope, key drivers, governance, planning mechanisms, commercial structure, operating model, and common pitfalls for these four basic models vary.

Other, related models of collaboration can be found in the useful left overs of the e-business literature, i.e. in Don Tapscott's *Business Webs.*

Neither competition nor collaboration are the straightforward answers to a lack of resources and an inability to exert control. Let me explain. Low collaboration and high competition lead to war-like behaviour, in which one party acts against others. Low competition and high collaboration typically lead to a situation in which partners integrate vertically with others. The best strategy for organisations is high collaboration and high competition, in a spirit of networking excellence, with the major benefit of competitive advantage (learning!) for all the participants.

Such collaborative networks of excellence occur best in situations where potential members have long-term recurrent relationships. Ideally, the organisations involved will conduct reciprocal, mutually supportive transactions. As organisations, such NoEs are neither markets, nor hierarchical firms. The participants retain their independence, while coordinating through negotiation and broad information exchange. Such collaboration leads to relationships of substantial trust, and a deepening confidence in each others capabilities and effort.

Successful networks of excellence combine the creative power of two or more entities and create a new core competence which would have been unattainable individually. With regard to resources, such collaborative networks expand capabilities, make possible economies of scale, outclass the competition through the establishment of de-facto standards, and reduce investment as well as financial and operative risks to the individual companies involved.

In essence, the ability of firms to collaborate in networks of excellence becomes part of their intrinsic value. A high presence of partnering and collaborative skills in a company carries its value far beyond that of its individual products and services, intellectual capital, and financial base. This becomes the true worth of companies in the networked economy: a blend of outstanding competences (Fig.2).

- Without collaborative skills, firms cannot participate in the global economy.
- Without elements of excellence, they cannot compete.
- Without a mixture of both in networks of excellence, they will not survive.

Fig.2: How firms secure their survival

Feasibility

Can we make networks of excellence work? The answers lie in a hard look at the risks inherent in collaboration, and at ways to offset such risks.

Firms enter into collaborative ventures not because they are the same, but because they are different! Perceived surfeits or deficits in knowledge, skills, technologies, core competencies and resources motivate companies to join in collaboration, and these differences remain essential for maintaining it. The gradual disappearance or convergence of these differences destabilizes the relationship.

For example, I was recently associated with two consortia. One consisted of companies with complementary skills, everyone contributed their expertise, and made a substantial contribution to the whole project to achieve something greater than the mere sum of its parts. The other had three heavyweight partners with near total overlap in their capabilities and skills. Every one of these latter partners invested considerable energy into edging the others out of the potentially most lucrative parts of the proposed project, and they never achieved the potential inherent in the bringing together of their three companies. They engaged in opportunistic behaviour and never achieved the level of trust needed for a successful joint approach to the project. They also lost the bid, and the considerable sums of money they had preinvested.

As we see from this example, trust is particularly important when collaboration takes place between competitors, and the risk of opportunistic behaviour is high.

Collaboration also has certain other disadvantages. Sharing with others can reduce the perceived level of management control. Increased dependence on external organisations can lead to greater need for co-ordination; this often translates into "more bureaucracy". Linking up closely may be perceived in the markets as a strategic alliance and will lead to both the opening and closing of doors, i.e. more or, in some cases, lesser access to other organisations and their capabilities.

There are also strategic risks. The relationship might finish, and one partner benefit from copying others. There is also the risk that one party might gain all the benefits from the venture. Therefore, risks and benefits sharing must be addressed and settled right from the start.

This is also true of the issue of costs, or, as some would put it, the business case to be made for each and every NoE. In forming and managing these joint undertakings, resources and transaction costs issues must be addressed right from the beginning.

A lack of understanding of each partner's motives in joining an NoE, for their operations, culture, and strategic intent can lead to resistance and conflict. Opportunistic behaviour, if it becomes the norm, destroys co-operation.

There is an inherent danger that – through a dulling of the competitive senses – partners lose their edge. More realistically still, partners may relinquish their competitive position by loss or transfer of their core competences, as a result of the perceived security of the new arrangements or, through rationalization pressures created by the NoE.

Proprietary information may be lost, the organisation may become disrupted, and conflict might destabilize existing business relationships.

There is a current perception that collaborative networks whose primary aim is to gain access to knowledge or ability are more likely to dissolve as internal skills and abilities are acquired through the partnership. Collaborative networks intended to gain benefit of scale or learning in performing an activity seem to have a more enduring purpose. A set of collaborators with complementary skills seems to be more enduring than a set with overlapping skills.

In order to make NoEs feasible, a number of success factors emerge.

NoEs need to create value for all participants, including all partners, their suppliers and their customers. The collaborators in an NoE need to focus on their key value drivers, such as efficiency, complementarities, lock-in, novelty and excellence. Success comes, among others, from optimising the value network among the participants, achieving time to market gains, the creation of effective governance mechanism (as yet, largely untried), measuring the progress and effectiveness of the network, continuous processing of inputs (such as environmental, market, customer knowledge, other expertise) into the network and the delivery of barriers which lock out competitors who are not part of the NoE, i.e. an element of competition directed at those outside the network. Difficult interfaces need to be avoided by adopting an open system model. Innovation and entrepreneurship need to be incorporated so as to propel the NoE forward. The boundaries between industries and their practices must essentially be spanned and blurred, so as to be able to define the most excellent practices in the value net.

There is a risky paradox, however. Networks can lead to an avoidance of taking responsibility, to evasive participation as part of an opportunistic set of behaviours: hoping to gain, but not willing to contribute, or only as willing to contribute as much as needed in order not to be kicked out of the collaborative venture.

Need for Accounting and NoE Governance

In order to establish an amount of management control, of intra-organisational relationships, and of self-governance of NoEs, two methods must be introduced and agreed among the participants.

One, accounting on a time and value basis, as the basis for a professional model that delivers the best value for all the partners. Second, the cultural design of a meaningful context of collaboration which keeps the value of collaboration in an NoE uppermost. The ideal is that outstanding participants and managers with reflexive capacities collaborate in the NoE, account for their time and act in accordance with cultural values designed for the specific task in hand.

To initiate, enforce and manage these desired outcomes, I propose a model of NoE Governance. This model, of course, needs greater refinement and will change as we gain more experience from working with NoEs. The challenge in this particular instance of governance is that accountability for it spans organisational boundaries. ewf europe suggests to act as the "honest broker", the neutral yet passionate institution to help establish such NoE Governance mechanisms.

NoE Governance is needed to make NoEs manageable, reduce the risks associated with them, improve their efficiency, and increase their impact on the bottom line of the participating entities.

The NoE Governance framework (Fig.3) contains the same elements as any corporate governance framework. Much like other frameworks, it is designed to ensure that NoEs work. It derives its additional complexity from facts such as

- unified business requirements cannot simply be deducted from an individual organisation's plan but have to be negotiated among the partners, along a methodology particularly suited to NoEs;
- Information confidentiality, integrity, availability, reliability, efficiency, effectivity, correctness need to be agreed on and constantly maintained;

- Planning and organisation (e.g. of data, applications, technology, equipment, personnel) is multilateral;
- Resources and their availability are an endless topic in most collaborative settings.

Controlling is a particularly challenging issue. In essence, controlling needs to safeguard management's capability to coordinate, react, adapt and assure the operative risks of NoEs.

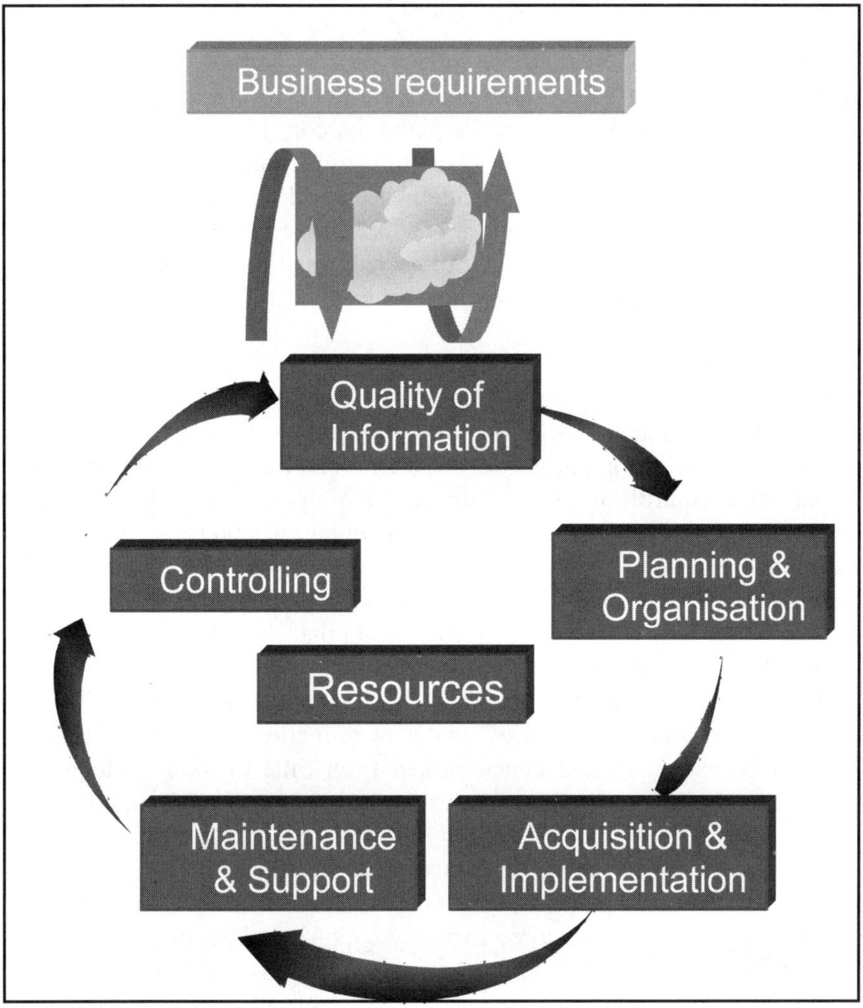

Fig.3: NoE Governance framework

Since NoEs will differ from each other in complexity, stability, duration, number of partners and strategic intent, it is difficult to make premature and fixed prescriptions.

However, four specific potential problems need to be (continuously) evaluated during the "run phase" in the context of NoE Governance:

- Selection of suitable partners
- Shared goals and values
- Evaluation of network benefits for the partners
- Collaboration rules and regulations.

An important task for both the "honest broker" and the controlling functions of the partnering firms is to make certain that systems are compatible, i.e. to create a tight fit between the planning, control and information systems of the members of the NoE. Traditional instruments for such tasks are "handbooks", i.e. collections of guidelines and rules.

Also, the accounting rules and procedures must be made to fit. Transfer pricing mechanisms, however, are often redefined for individual networks and may differ from practices in other business relationships of one or several participants in the NoE.

A word of warning: the search in NoEs is for complementarities first, not for financial gain. Therefore, qualitative measurements must be used to capture the consequences of either opportunistic or cooperative behaviour.

As to controlling goals, profit considerations play a paramount role in the in-house controlling of any firm. But in NoEs, this is complemented in importance by other considerations, for example, whether the competences assembled and/or the resources provided are sufficient to assure the long term success and expansion of collaboration. Similarly, an early warning system needs to be installed to make certain that the allocation of tasks is seen as fair, and that preferential treatment is avoided. The latter would be an indication of structural imbalances leading, as a rule, to destabilization and possible failure of the NoE under examination. Transfer prices also need to be examined and benchmarked from time to time, in long term NoEs. Simulation models have been found useful for these complex tasks.

Top managements must, of course, take care that NoEs take into account the regulatory environment, be that Basel II, the management accountability rules now in force in many countries, Sarbanes-Oxley in the United States, or the validation requirements of many industries.

Sustainability

The economic shocks of the past three years – terrorism, exploding market bubbles, rapid shifts in the regulatory environment, technological and market disruptions – all require that hard questions be asked, throughout their lifecycles, if NoEs are to be sustained, and become lasting successes.

What are the environmental preconditions, the factors which persuade us that an NoE has a likelihood of success? When are collaborative relationships such as those in an NoE appropriate? What types of collaborative relationships exist, and which is appropriate under which conditions? How do you create and sustain these relationships?

Environment	1. A History of collaboration or cooperation in the community 2. Collaborative group seen as a legitimate leader in the community 3. Favourable political and social climate
Membership Characteristics	4. Mutual respect, understanding and trust 5. Appropriate cross section of members 6. Members see collaboration as in their self-interest 7. Ability to compromise
Process and Structure	8. Members share a stake in process and outcome 9. Multiple layers of participation 10. Flexibility 11. Development of clear roles and policy guidelines 12. Adaptability 13. Appropriate pace of development
Communication	14. Open and frequent communication 15. Established informal relationships and communication links
Purpose	16. Concrete, attainable goals and objectives 17. Shared vision 18. Unique purpose
Resources	19. Sufficient funds, staff, materials, and time 20. Skilled leadership

Fig.4: Wilder success factors for collaboration

The Amherst H. Wilder Foundation (*www.wilder.org*) of St. Paul, Minnesota, has been looking at collaboration as a means to improve communities for the last one hundred years. Its first report on collaboration appeared in 1915. Its Wilder Collaboration Factors Inventory features twenty success factors for collaboration, in six groups (Fig.4).

Collaboration does not always constitute the best way to accomplish tasks or address issues. If collaboration in NoEs is to achieve something of value, we must take into account the experiences already made elsewhere. The Wilder inventory, for example, gives us a basis for a quick SWOT analysis, as well as for renewal if we encounter problems in the running of NoEs.

The NoE teams themselves are very important. A sufficient body of research exists to help with the establishment and development of the cross-functional teams needed. As a first step, the teams in NoEs need to be well designed, typical project roles and responsibilities need to be defined, and they need to be equipped with a team charter. Every effort must be made to build efficiency into their work, and help must be provided to manage the inevitable conflicts. Steering teams need to assure the necessary strategic alignment over the life cycle of the NoE. Inevitable conflicts of interest must be identified, and dealt with through constructive strategic dialogue.

Of particular importance – and recognized in the research literature – is the need to have a formal project kick-off. Here's a reminder of what such a kick-off could involve:

1. Review plan
2. Review timeline
3. Demonstrate executive sponsorship
4. Define level of commitment
5. Instill confidence
6. Describe planned training for team members
7. Demonstrate products for collaboration, if used
8. Specify deliverables by phase
9. Define roles and responsibilities
10. Introduce project team.

At ewf europe, we look forward to helping with systematic and enthusiastic kick-offs in all our overall NoE programmes, as a first step towards the success of individual NoEs. A successful kick-off is always an opportunity to celebrate, and – if nothing else – we want to be there, too!

We know that NoEs are uncharted territory. We will establish substantial programme management capabilities to address the major shortcomings which currently exist in dealing with NoE life cycles, in innovative cost and financing considerations, and in the opportunities and risk management inherent in the models.

ewf's programme management will play to the strengths of the NoE concept and deliver overall solution to the known problems: all of this intended to make NoEs sustainable (Fig.5).

Programme management:	Examples of key words
Optimise Process Impacts	Coaching, facilitation, research, collaboration, multidisciplinarity, cross-disciplinarity, training, partnerships
Enhance Knowledge Transfer	Effective ways, dissemination, cross-fertilization
Enhance Technology Transfer	Effective ways, dissemination, cross-fertilization
Protect Intellectual Property	Protection vs. exploitation, context-specific
Achieve Long-term Goals	Renew, redirect, support, coaching, crisis intervention, conflict settlement; measurements (quantitative / qualitative), productivity and economic growth

Fig.5: Tasks for NoE programme management

Conclusion

In this article, I put NoEs through three tests: relevance, feasibility, sustainability.

The exploration of these tests and their implications shows tremendous opportunities for NoEs, but also implies costs and risks associated with such undertakings.

If we want to avail ourselves of the opportunities associated with NoEs, we must aim to have a wide scope, achieve as wide a frame of reference as possible. This involves a systematic method of development, including

- a look at the broad landscape of existing collaboration within and among industries, and at our own personal possibilities;
- a vision of the best possible outcomes, goals, or solutions for NoEs in order to become energized;
- flexible yet determined plans for the delivery of our NoE vision, and – this is important – giving up the counterproductive behaviours which inhibit collaboration;
- enlistment of others in order to gain energy and support;
- management of progress and driving the momentum;

• dissemination of our successes and achievements.

If we want to make certain to manage the costs and risks associated with NoEs, we must avail ourselves of the very best NoE Governance ideas and practices, as they develop.

Finally, let me take you some way into the future. As of now, we are only at the beginning of NoE practice, and certainly of NoE theory. Over time, I would expect a network theory to develop which would help us to apply refined methods and procedures in order to manage opportunities, costs and risks in the networked economy even better. Every time significant NoEs happen, they will provide new data, new knowledge, and contribute to our understanding of the models.

At an advanced stage, we shall be able to develop to a point where we can suggest which NoEs to connect to each other, i.e. we will span networks such as we span enterprises today.

Network theory will take off if we keep our work multidisciplinary, comparative and international in nature. I am glad for and excited by the personal opportunity to do so, and look forward to working on our perennial challenge: create value, wealth and new opportunities for everybody involved in NoEs, while reducing risks and costs.

References

Stephen Cardell: Strategic Collaboration. Creating the extended organization. Abingdon: Hodder & Stoughton, 2002

John Child, David Faulkner: Strategies of Co-operation. Oxford: Oxford University Press, 1998

Mark Ebers (ed.): The Formation of Inter-organizational Networks. Oxford: Oxford University Press, 1999

David Faulkner, Mark de Rond (eds.): Cooperative Strategy, Business and Organizational Issues. Oxford: Oxford University Press, 2000

Henry J. Lindberg: The Basics of Cross-Functional Teams. New York: Quality Resources, 1997

Tom Peters, Robert Waterman: In Search of Excellence. New York: Collins & Row, 1982

Paul W. Mattessich et al.: Collaboration. What makes it work. A review of research literature on factors influencing successful collaboration. St. Paul: Amherst H. Wilder Foundation, 2001

Walter J. Michalski: 40 Tools for Cross-Functional Teams. Building synergy for breakthrough creativity. New York: Productivity Press, 1998

Peter Sachsenmeier: Dot.coms: Putschmittel für Finanzwelt, Kultur und Technologien? In: Ulrich Siebert, Jens Meyer: Blenderwirtschaft. Frankfurt: Frankfurter Allgemeine Buch, 2002, pp. 58-76

Don Tapscott, David Ticoll, Alex Lowy: Digital Capital. Harnessing the Power of Business Webs. London: Nicolas Brealey Publishing, 2000

Secure Engineering Collaboration

Engineering Alliances among Collaborative Enterprises

Martin Schmidt

Director Business Center, Engineering Process and Data Management Solutions, T-Systems ITS GmbH

Through the globalisation of markets and technical innovations engineering of complex industrial products has changed dramatically through the last years. Products life cycle is getting shorter and at the same time customers are expecting innovative products for an attractive price.

Companies have detected that they are only able to react on these requirements if they get their engineers highly productive and if they cooperate with best in class partners to get access to innovative ideas as well as solutions at the right time.

In the last years companies introduced Product Data Management Systems into their companies as well as intelligent product design applications to support their engineers. To get control over the engineering process workflow solutions and project management tools where introduced. To support collaboration with other engineering teams, companies introduced data exchange solutions and various communication tools to synchronize their engineering work.

Nevertheless a lot of companies do not feel comfortable with their approach in engineering collaboration. Companies are missing standards; investments into IT solutions are high; security issues are unsolved and the management of processes across companies is not well defined.

For this reason T-Systems worked on a concept and services with customers to enable real secure engineering collaboration in the future. The concept named “eBridge” includes:

- Templates for engineering collaboration (organization forms, definition of roles and responsibilities, project templates; proofed standards, etc.)
- Management of engineering projects across companies
- A set of predefined IT/Telecommunication services to support engineering collaboration.

Within the workshop T-Systems presented the eBridge concept and an overview about IT/Telecommunication services to support engineering collaboration.

e-Business: The Importance of Standards for e-Marketplaces

Alexander Scholz

Nexolab GmbH

Preface

The success of the Internet, which is the basis of today's e-Business, was only possible because TCP/IP was acknowledged as the foundation of communication and data exchange. This rapid success of the Internet has changed the way we used to conduct business, but the first experiences that have been made, demonstrated that this new technology requires a new understanding of how to interact with one another.

Standardisation has always been a very controversial field of activity as the benefits of standards simultaneously reduce the possibilities of individuality and individual representation.

This chapter is not a prayer for standardisation, it is rather the attempt to identify areas that can be successfully standardised without loosing important individuality. Especially for e-marketplaces that act as intermediaries for a wide range of enterprises and organisations the matter seems more than important.

Based on this knowledge this forum has a double function. On the one hand, it seeks to define possible solutions for the areas that are covered and presents these solutions as recommendations to the EU Commission to be implemented within the 6th EU Framework. On the other hand, this forum is the stating point for the 'Standardisation Network of Excellence', which means that this dialogue needs to be continued on a regular and international basis.

Summary

This section of the book clearly focuses on the Dos and Don'ts of e-Business and tries to elaborate activities to overcome the current difficulties that there are in e-Business. Based on the experiences of the last three years the authors of this section have selected the most important areas of standardisation. They have furthermore analysed the current situation and the most important challenges.

The first section deals with electronic catalogues. Electronic catalogues play an important role as they are the basis for a standardised description of products and services.

Peter Mund, Business Development Manager at the BMEnet and Project Manager for the European Content Club (ECC) and the European Content Factory (ECF) is highlighting the current situation on electronic catalogues and is pointing towards possible solutions and recommendations.

The second section picks-up transaction and transaction standards as they are key to data exchange between entities and as they enable seamless digital business process.

Boris Otto, project manager for electronic marketplaces for the Fraunhofer Institution for Work Science and Organisation (IAO) is highlighting the current situation of transactions and transaction standards and is depicting possible solutions.

The third section deals with business processes as most e-Business saving potentials are a result of leaner and more standardised and thus more integrated business process.

Mr Jacques Spee, Director Solution Consulting at EDS in the Netherlands has selected an interesting example in order to highlight both the current status and the challenges with regards to business process standardisation. He also points towards potential solutions and recommendations.

The fourth section deals with a generic reference model as an example of future inter-enterprise collaboration. Udo Döbrich from Siemens Automation and Drives further highlights the importance of common product data.

Introduction

The New Requirements of Networked Economy

Current e-Business needs to evolve from what it offers today. The technological foundation of the Internet has enabled enterprises to communicate with each other. Inter-enterprise communication defines new challenges for the way we are doing business. Future requirements make it inevitable that a more integrated e-Business enables what was promised three years ago. Some new requirements are the following:

1. Seamless integration of inter-enterprise business processes
2. Central data repositories and centralised data storage
3. Automated data exchange by reduced interfaces
4. Collaborative business processes

Importance of Standards for an Integrated e-Business

The Internet became successful as it enabled the common use of standardised technology. It also enabled cheap access to technology and supported rapid implementation. These success factors were the foundation for the fast conquest. After the e-Business hype in the end of the Nineties, and the current frustration, the Internet needs to expand the technological standards towards a standardisation of content. In other words the Internet needs to evolve towards an integrated e-Business platform.

Major stepping stone for this integration is the deployment of common standards, as e-Business potentials can only be realised if processes are automated and automation requires a common language.

The Five Dimensions of Standardisation

When we talk about standardisation, we talk about five different dimensions that need further standardisation as they are central elements of an integrated e-Business.

1. Electronic Catalogues
2. Classification
3. Transactions Protocols
4. Technology
5. Business Process

Definition of Terms

Many terms that are common knowledge in today's e-Business are many times used in a confusing manner. In order to avoid this confusion, we briefly define the terms as we have applied them during the forum. The definition also helps to understand the scope of activities that has been kicked-off during the forum.

Electronic Catalogues: Electronic catalogues are the basic format for describing both products and services.

Classification: Classifications are the structure behind each catalogue (electronic and paper-based) grouping products and services.

Transaction Standards: Transaction standards are the formats that define the content– both required and additional - of a business transaction between two or more entities.

Technology: Technology is the underlying enabler of inter-enterprise communication

Business Processes: Business processes are a sequence of transactions between two or more entities.

The Horizontal Approach

In order to fully cover the different aspects of standardisation, we have chosen to involve five different interest groups each with a dedicated role.

1. Associations aggregate the requirements of an industry
2. Marketplaces develop de-facto standards by enabling many-to-many relationships
3. Research Institutions think forward define new processes and standards
4. Enterprises need to participate in different portals and marketplaces
5. Consultancies know the challenges in a wide range of enterprises

Electronic Catalogues

Peter Mund

BMEnet, Business Development

Definition of Terms

Electronic catalogues are the central place, where information about products or services/support is administrated. Therefore electronic catalogues should contain all necessary and relevant product and service information both for buyer and supplier. The following fields are the minimum requirements for any electronic catalogue:

1. Order code
2. Product code
3. Picture, drawing, layout etc.
4. Price per item
5. Price per quantity
6. Minimum order quantity
7. Manufacturer
8. Supplier
9. Delivery time
10. Short description
11. Long description

Inside the electronic catalogue the user is not able to place any order. The user chooses relevant products and places them into the product-basket. After the selection procedure the information will be passed to the next level procurement system.

Types of Catalogues

The definition of a catalogue can be opened in principle as a function of the product characteristics over three sources (Fig.1).

Purchasing organizations often face the challenge to decide for the catalogues that are going to deployed in their Intranet. This decision process causes serious delays in the project progress and thus substantially rises the costs of the project.

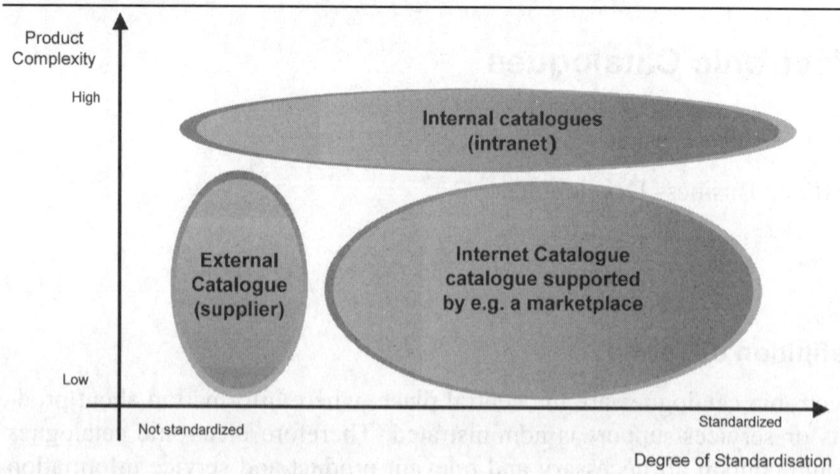

Fig.1: Different types of catalogues

Single Supplier Catalogue

In case of single supplier catalogues every supplier puts his own catalogue into the catalogue management system.

In contrast to multi-vendor catalogue, the eProcurement system can use more than one single supplier catalogue.

In order to avoid maintenance problems a uniform standard should be deployed in catalogue management.

Multi-vendor Catalogue

The multi-vendor catalogue contains the data of more than one supplier. If identical products are offered by several suppliers, then the multi-vendor catalogue allows a direct comparison of the products, e.g. price, technical features.

A general condition is that the multi-vendor catalogue is hosted at a central environment. A uniform look and feel and a standardized product classification enable the most effective use of the multi-vendor catalogue.

Critical Success Factors for e-Catalogue Management

Data Exchange Formats and Classifying Schemes

Data exchange formats enable the workflow-based exchange of catalogue data from supplier to buyer and are thus basis for updated data. Data exchange formats have to be constructed such, that users are able to locate required products quickly.

Four success factor are determining the user friendliness of the each e-catalogue, these are the following:

1. Standardisation,
2. Classification,
3. Rationalisation and
4. Normalisation.

Standardisation

Standardisation means the deployment of a catalogue standard that regulates the electronic transmission of product data and contains a catalogue classifying scheme.

1. Generally accepted and category-specific attributes
2. Classifications/categories for the products
    - BMEcat
    - cXML
    - xCBL

Classification

Classification of products in product groups/categories requires a lot of industry knowledge. There are standards that allow suppliers and buyers to agree on categorisation. These standard classifications are essentially a long list of categories and corresponding standard article numbers that assign articles to each category, current classification schemes are

1. eCl@ss
2. UN-SPSC
3. ETIM

The chapter on classification further highlights important details.

Normalisation

Normalisation is the common use of terminology for supplier specific abbreviations and terms. This also includes supplier special characters.

Localisation

Localisation is another important success factor of catalogue data as it adjusts data according to national defaults, e.g. the use commas instead of points for numerical subjects.

Rationalisation

Sequence of the article indications and the article characteristics after their meaning.

Data Format

Uniform data formats for the setting up of catalogues and the exchange of product data within a procurement system and the electronic catalogue, e.g. XML. To find more information, see chapter *Transaction Standards*.

Conclusion

A penetration of industrial-specific standards for the catalogue exchange functions is absolutely to regard. Therefore it is necessarily to set up international standards.

The German BME, the association for materials management, purchasing and logistics, is the leading association for purchasing and logistics in Germany.

The BME is the main engine driving the ongoing development of e-Business transformation (e-Standards, e.g. *BMEcat and OpenTrans*). BMEcat, pushed and promoted by purchasing managers of leading major corporations in Germany, is the most widely used standard for the exchange of multi-media electronic product catalogues in German-speaking countries. www.bme.de

Conclusion

In order to implement the above mentioned success factors, we propose the following steps

- identify similar organisations across Europe with focus on this standardisations
- define responsibilities inside these organisations
- establish a working group across this organisations for the promotion of this standard
- define projects to promote the standard
- introducing the standard to further organisations

These steps should be started as soon as possible including a both national and EU Commission representatives. A potential industry project could help to verify the results of the working groups and to foster credibility of the common efforts.

Transaction Standards

Boris Otto

Fraunhofer Institution for Work Science and Organisation (IAO)

Definition of Terms

Since electronic business applications mainly aim to reduce costs of conducting business between two or more business partners, the role of business transactions is critical for the overall success of any electronic business engagement. To achieve efficient inter-company business processes integration is the key. Integration means both system integration on an application level and also process integration on a business level. The linking element of both is a business transaction. In general terms, a business transaction is defined as a relationship between business objects that works as an interaction channel between two of them.[1] A business object represents e.g. a supplier, a buyer, or a business document. Apart from that, the term "transaction" is widely known in the field of database theory as a database operation that follows the ACID rule, i.e. the operation is compliant with the principles of atomicity, consistency, isolation and durability.[2] In recent years, the latter interpretation of the term is also made use of in the area of business transactions because database requirements turned out to be applicable to business transactions, too. In case a buying organizations transfers a purchase order via HTTPS, it must be made sure that the document exchange takes place completely or is subject to a roll-back routine in case it did not.

As mentioned in the beginning, business transactions form the backbone of every efficient electronic business process because it links the system integration with the process integrations perspective of the subject matter. Due to its enormous impact on working electronic business processes and due to the numberless different trading partners, applications, operating systems etc. involved, it becomes clear that standardization in the field of

---

[1] Rump, Frank J., Geschäftsprozessmanagement auf Basis ereignisgesteuerter Prozessketten, Stuttgart, B.G. Teubner, 1999

[2] Gray, Jim, Reuter, Andreas, Transaction Processing – Concepts and Techniques, San Francisco, Morgan Kaufmann, 2000

business transactions is a critical success factor for every electronic business engagement.

Keeping this in mind, the following chapter gives an overview on the layers of business transactions on which standardization is needed and on the work that is currently being carried out in the related fields.

Reference Layers of Business Transactions

Talking about business transactions, the following different layers of information exchange can be identified:

1. The Protocol Layer
2. The Business Document Layer
3. The Process Framework Layer
4. The Web Services Layer

The Protocol Layer forms the foundation of electronic business communication and information interchange. It mainly defines the technical circumstances under which the business transaction is carried out. Among these pieces of information are e.g. identification parameters of the sender and the recipient. Therefore, the Protocol Layer enables network communication for distributed environments. Based on that, the Business Document Layer specifies the so-called payload that is exchanged within a business transaction. On the Business Document Layer transaction or document standards specify the messages that are exchanged between one or more companies. Among the most relevant business messages are purchase orders, invoices, dispatch notifications etc. Since business document standards only specify the message and not the order and sequence in which the messages are exchanged, Process Frameworks specify the business process on top of simple document exchange.[3] However, Process Frameworks mainly aim to facilitate stable and deterministic business processes. In order to also automate and facilitate dynamic process environments, Web Services are needed that describe certain business functionality as end points within a dynamic network.

[3] Frank, Ulrich, Standardisierungsvorhaben zur Unterstützung des elektronischen Handels: Überblick über anwendungsnahe Ansätze, Wirtschaftsinformatik 43 (2001) 3, p. 283-293

Current Standardization Initiatives in The Field of Business Transactions

The following table depicts current initiatives in the fields specified above.

Table: Current Standardization Initiatives

Protocol Layer	Business Document Layer	Process Framework Layer	Web Services Layer
FTP	BMEcat	eCo	BPEL4WS
HTTP	cXML	ebXML	WSDL
SMTP	openTRANS	RosettaNet	UDDI
SOAP	ANSI X12		
	xCBL		
	DIN 5 XML		
	EDIFACT		

With regard to the Protocol Layer, most of the standardization work has been done during the last 10 years. As a result a working set of different protocols for different forms of communication is available. FTP (File Transfer Protocol) is used to exchange large amounts of data, HTTP (Hyper Text Transfer Protocol) is mainly used for synchronous and time critical information interchange whereas SMTP (Simple Mail Transfer Protocol) is used for email. Rather new is the SOAP (Simple Object Access Protocol) approach that works as a remote procedure call for the web.

Various approaches exist in the area of business document standards. It can be distinguished between "classic" EDI approaches such as EDIFACT or ANSI X12. These are used for long years for very lasting stable business relationships between business partners. They are very well suited for the exchange of high-volume data. Other initiatives are based on the XML specification, e.g. xCBL and openTRANS. These formats make use of the advantages of XML and are therefore well suited for flexible, web based applications. When deciding which technology to implement (XML based vs. classic EDI), a set of evaluation parameters have to be taken into account such as data volume, network availability, software support, user-friendliness, message set, target group (human being or machine).

Process Frameworks can be divided into two groups, namely horizontal approaches and vertical approaches. Horizontal initiatives do not target a certain industry or domain of application both rather try to provide a generic approach that can be used across industries. The most popular horizontal process framework is ebXML (electronic business XML) that is currently maintained by UN/CEFACT (the business related parts) and OASIS (the technical parts). An example for a successful vertical initiative is RosettaNet that specifies business processes for three industries, namely the semiconductor manufacturers, the electronic component industry and the information technology industry. RosettaNet is now subsidiary of the Uniform Code Council (UCC).

Mentioning Web Services, the so-called stack is considered quite stable. Based on SOAP, WSDL is used to describe business functionality as a service end point that is accessible via standard web protocols. A service provider can then publish the service in a service registry. Such a registry can be standardized using UDDI (Universal Description Discovery and Integration). A registry can be used by interested parties to look up a certain service and then automatically send a request to the service provider. Specifications such as BPEL4WS (Business Process Execution Language for Web Services) can then be used to define "choreographies" of single web services, thus modelling entire business processes out of web services.

Conclusion

The field of business transactions can be divided into four layers. For every layer various standardization initiatives can be identified. Taking this into consideration, future research must focus on the improvement and applicability of existing standardization initiatives rather than on the development of new ones.

Applied research must answer the following strategic questions, in the short term:

How can applicability of standards be improved?

How can utilization guidelines be made available to the user community?

How can administrative matters such as legal and governmental aspects be included?

How does an overall set of electronic business standards look like?

How can technological risks of investment be minimized?

Which approach covers semantic problems and not only syntactical matters?

Business Process Standards

Jacques Spee

Director Solution Consulting, EDS

Collaboration between companies is a critical success factor in most industrial manufacturing markets. The value of electronic business is to enable large efficiency gains and more responsiveness in manufacturing-supplier networks by adopting web technology to support key processes. Essential for co-operation in an e-Business environment are aligned business processes, or rather interoperable processes across the internet, hence the need for e-Business process standards.

Recent Development & Current Status

The best known electronic business solutions are marketplaces for business-to-business (B2B) networks in specific industries sectors. Examples of e-Markets are Odette (automotive), Exostar (aerospace), RosettaNet (electronics), and Covisint (automotive). Typically these e-Markets aim at cost reduction and speed increase in supply chains.

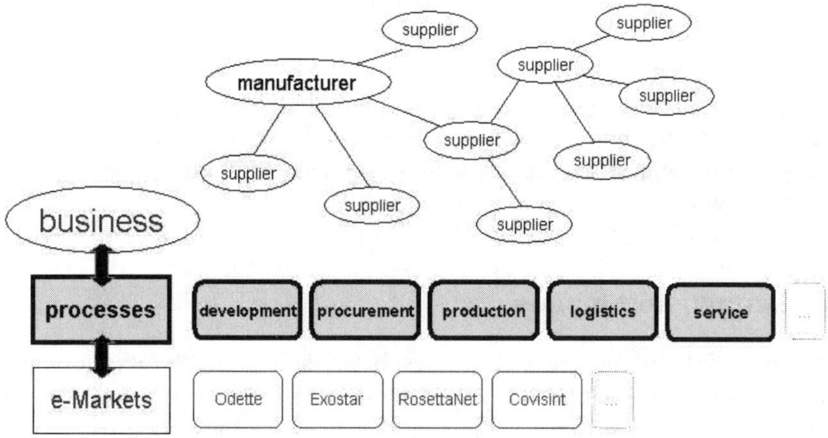

Fig.1: Overview of business processes and e-marketplaces

Most e-Markets originated as electronic catalogues, evolved into trading places for components and materials, and then integrated supply chain management functions. The latest addition is to include product development and service/support in the e-Business scope for B2B. In the current trend also customer relations and sales are considered viable parts of e-Business, thus covering almost the complete product lifecycle.

Current e-Markets provide a framework for electronic communication between supply chain partners by means of standardised transactions. Examples of this are Covisint's Solutions Suites, RosettaNet's Partner Interface Processes, and Odette's so-called Message Sets. These standards provide message interchange protocols and document formatting to ensure correct handling for specific inter-partner communications.

Involved Organisations, Institutions and Firms

The organisations that primarily drive e-Markets are Original Equipment Manufacturers and Information Technology vendors. Some e-Markets originated from Tier-1 suppliers, like SupplyOn in the automotive segment. The management of these markets is often organised in separate firms, that sell their solution in specific markets and also facilitate further development of the transaction framework with the help of subject matter experts. The effect is a multitude of standards for various market segments, where each e-Market firm is striving to become the 'de-facto standard'.

Several initiatives for the development of 'open standards' in e-Business are run under auspices of standardisation bodies, with involvement from both public and private parties. The International Standardization Organization has a range of active working groups in industrial technical committees. The United Nations is central to renewed interest for existing Electronic Data Interchange standards, like the updated UN Standard Messages. In the area of internet standards the World Wide Web Consortium develops and promotes vendor-independent standards, of which XML is best known.

Challenges and Issues

Interoperability of business processes between manufacturers, suppliers, and customers is essential to reap the full benefits of e-Business, and to enable its use across entire value chains. Current e-Markets provide standardised transactions as a key component for *data exchange* in electronic business. The next step is to standardise B2B components for *process cou-*

pling, such as process description languages, workflow execution services, and robust exception handling, in order to achieve integrated e-Business processes in manufacturing networks.

A combination of business and technical challenges and issues remain for widespread e-Business process deployment and application:

- *Cost & Value* – Recent years have shown that the investments required by organizations to integrate e-Business processes and IT systems is quite high. Funding in the current economic climate is restricted, and this leads to redirection of investments to high value business processes that align with business strategy. Improved customer relations and streamlined internal operations are expected focus areas for investments, with lower priority on supply chain spending.
- *Process Innovation* – business processes are considered a competitive factor by many companies, and e-B2B would make processes transparent to other parties. Current transaction frameworks in e-Markets give only limited visibility of process state to the trading partners. The purpose of new e-Business process standards is to be able to integrate automated processes, not only inside one enterprise but also across the value chain. Collaboration agreements and technical provisions are necessary to enable streamlined and agile intercompany process coupling, and at the same time protect intellectual property in business process capabilities.
- *Connectivity* – Web Services is the latest technology that aims to provide open standards for applications interfaces and data formats, thereby facilitating the integration and automation of e-Business processes. Manufacturers and suppliers alike are generally not fully convinced of the business proposition for electronic B2B process integration across the supply chain using Web Services. An overall vision and proven benefits for manufacturing industry application are not yet present.

Recommendations

The potential benefits of e-Business process standards in the manufacturing industry remain high, but further development is needed to prove the business value and demonstrate successful implementations to stimulate broad adoption. These subjects fit very well in the European Commission's sixth Framework Programme, for which the following recommendations are made:

- Start a *European Network to develop e-Business Process Standards* for manufacturing industries. The Network objectives are to define reference process models using standardised terminology and classification structures, and to ensure European influence in definition of these critical e-Business components. The Network should represent manufacturers, suppliers, industry associations, standardization bodies, manufacturing researchers, and technology providers.
- Facilitate and sponsor *Pilots Projects for e-Business Process Standards* in manufacturing value chains. The Pilot objectives are to prove benefits of e-Business process integration for various stakeholders, to validate e-Business process standards, to test standards-compliant technology, and to disseminate proven solutions across manufacturing industries in Europe. The Pilots participants should represent manufacturers, suppliers, standardization bodies, and technology providers.
- Adopt a *European e-Business Standards Organisation*. The EU Standards Organisation objectives are to facilitate the development of vendor-independent e-Business standards, to formally own and distribute such EU e-Business standards, and to certify organisations and products to comply with these EU e-Business standards. The European e-Business Standards Organisation should represent the EC, national and industry standardization bodies, and industry associations.

A Generic Reference Model for the Description of Plants and Products

Udo Döbrich, Roland Heidel

Siemens Automation and Drives, Karlsruhe

Information Technology is about to enter all company related processes. Associated Data will represent more and more the "contents of business" such as development, manufacturing, marketing and financials.

After years of struggling with bunches of different tools created by uncoordinated design approaches for sub applications it is now obvious that the application "Company" has to be considered as a system consisting of different processes which have to be supported by consistent tool chains.

The backbone of this approach must be the seamless exchange of information between all participating processes (Fig.1).

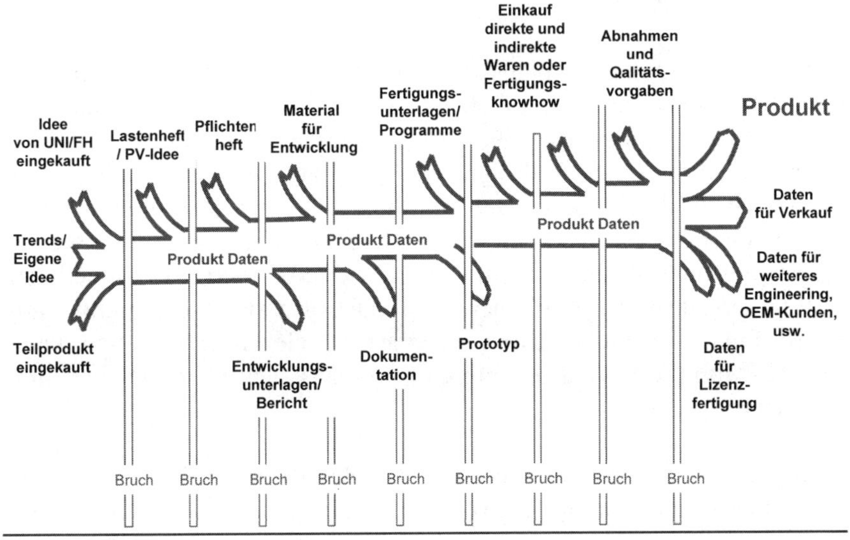

Fig.1: Product data flow in a company

For this purpose the definition of an overall product data reference model is needed to allow describing the semantics of process data in a company. The reference model brings together the different existing process islands into a transparent world of information similar to the OSI seven layer model for open communications.

Any product is characterized by data called attributes describing the product properties. These data are the basis for any selection of a product i.e. from an electronic market place by a customer. Nobody buys a bulb without having a look upon its technical attributes like shape, connector type, voltage, etc. Whilst commodity items like bulbs usually have only a few attributes the products in industrial automation have several hundreds. Therefore it is necessary to define **basic elements** for the classification of attributes such as:

C = for Constructional attributes

F = for Functional attributes

P = for Properties attributes

L = for Location attributes

B = for Business attributes

The constructional attributes describe the mechanical or constructive data of a product. The functional attributes represent the functions of a product, which can be either functions in software or hardware. A car has constructional attributes as design, functional attributes are driving, stopping, change of direction etc. But what type is the fuel consumption or the acceleration time between zeros to 100 km/h? These attributes are the property attributes of the car. They are neither constructive nor functional. Location attributes describe where a product or parts of it are physically located. Finally business attributes contain all relevant information which has to do with the buying or selling of products such price, delivery time and others.

But these basic elements are not just a classification; they provide the description of interrelations between each other in an assembly of several items (Fig.2).

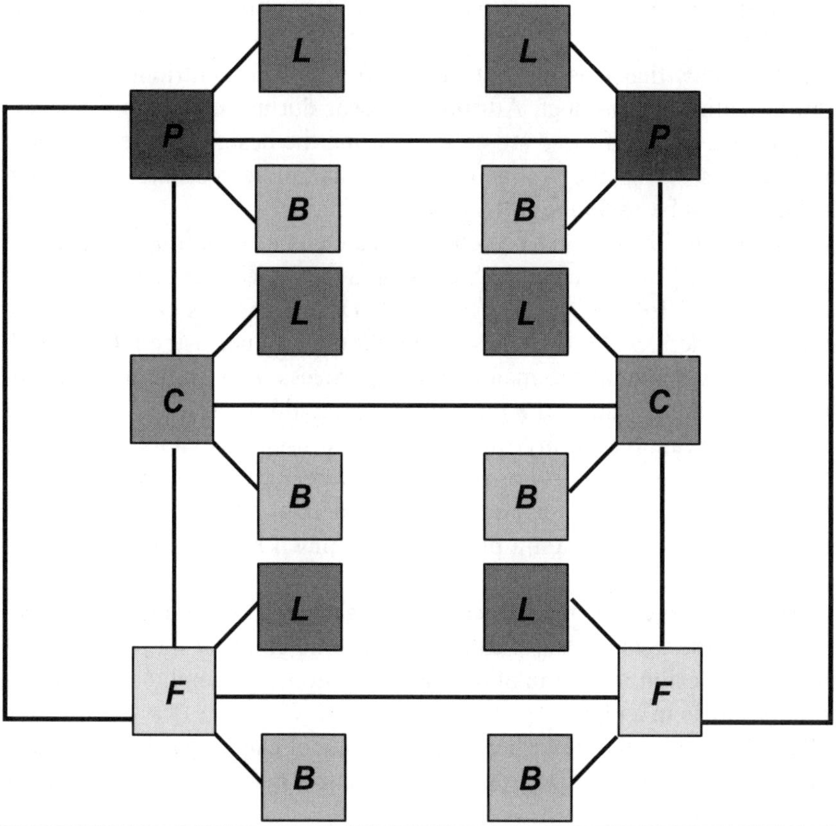

Fig.2: Interrelationship between basic elements

Constructional elements may be connected to each other. For example an electronic board is plugged into a rack or two devices are connected via a communication system. Each constructional element owns business elements like price. There is an interrelation between the constructional attribute and the business attribute. Constructional elements also have an interrelation with location information describing for example the position of an electronic board in a rack. Of course functions have interrelations, too. One function exchanges information with another function. Functions also have an interrelation with constructional elements, for example a software function runs on a PC. Finally the PC has property attributes like the amount of memory, clock frequency or performance of the software.

These examples may give some feeling for the interrelations between the basic attributes of products. Not all these interrelations must be present for a product at any time, some may be necessary only in certain periods of the product life cycle.

It is obvious that it is inefficient just to write these attributes down in front of selling the product. Attributes appear during the whole live cycle of a product. Some of these are created during the design and development phase, others during the production or for marketing reasons. All created attributes must be usable for all the various purposes.

An example for the flow of constructive element data in a manufacturing process makes this clearer. A constructive part can be developed by a method like STEP (ISO 10303/Express). During the design process, the developer generates the attributes including their values using a CAD tool. After the development the manufacturing process adds some more attributes with specific values, the marketing people do the same. All these attributes are available at any time and can be used to extract the needed technical and commercial properties for different purposes. A buyer only needs a small amount of technical and commercial attributes whilst an engineer who wants to design a plant requires much more detailed information.

Based upon the systematic electronic description of product data it becomes possible to apply these data for electronic engineering processes including the design, the simulation and validation of a combination of different products in a plant.

This shows that the described reference model can be used as a generic method not only for the description of products but even for the description of a complete plant.

It seems that this reference model can also be used to describe personal related and organisational aspects within a company respectively in any process. Thus this *CFPLB Reference Model* originally designed for usage in the area of Industrial Automation should also be appropriate to describe any *Distributed Company* and its related aspects (iPOT).

Conclusion

Roland E Wolff

Nexolab GmbH

Standardization in e-Business is the vehicle for the industry to produce and direct mass-customized products and information for the consumer. It is a paradigm in itself but it's the only way if e-Business will finally create the value which has been promised since the beginning of the Internet. The wild years of e-Business have produced various ways of using that new media at a mushroom growing rate. Its is now the time, after a huge amount of money had been available to produce that wide array of different languages, transaction (codes), catalogue formats, etc. to consolidate and filter out the ones that could be used as a foundation for a solid development of standards for the e-Business community.

In those sessions described before the teams have concluded that there is a need to look at the different aspects of e-Business

1. On a European level
2. Identify the current status of activities
3. Develop a future state design
4. Identify projects to be approved by the EU as Integrated Project within the 6th EU Framework.
5. Implement rapid projects within the Framework of the 6th EU

The current world economical problems will increase the pressure on companies to improve their productivity at even faster rates. Standardized e-Business behaviour eventually can be the key to a success story everybody has hoped for. Unlike in the hype days realism is now the name of the game and lots of energy is needed to structure the before discussed items.

Neutrality, Openness and Co-operation in the Service Industry

Hagen Wenzek, Harry Teifel

Hagen Wenzek, IBM, Harry Teifel, Dekra AG

Introduction

Basically, collaboration and competition are historically well known aspects of economic behaviour. In Germany, they did come as associations, guilds, and collectives. These entities had been established to adjust imbalances between economical partners of different sizes.

That approach institutionalised today's concept of co-operation between competitors and of collaboration. The uneven distribution of power between a major manufacturer and its suppliers or service providers is meant to be equalized. Despite being competitors on the marketplace, the joining parties would be better positioned in their relation with the major manufacturer, than by staying on their own.

However so far, this form of collaboration has mainly been an organizational phenomenon. Only since the widespread availability of IT support for these networks it became possible to seamlessly integrate processes as well. With such an infrastructure in place it is feasible to drastically enhance productivity through collaboration.

Frictions throughout a Product Lifecycle

Within the lifecycle of an industrial product, we find several potential inhibitors of seamless cooperation. Taking e.g. the automotive lifecycle starting at *Research and Production*, we see relations between:

- OEMs and suppliers
- OEMs and other OEMs
- Governments and OEMs

When, e.g., the OEM puts highest pressure on the margins of the supplier, frictions between both players extend and the relation departs from a true collaboration to a pure power play.

Going further to *Sales, Distribution and Customer Support*, other partners come into play extending relations to:

- OEM, importer and retailer
- End-customer and OEM and/or retailer

As an example, a high degree of collaboration is established through the legal construct of the Group Exemption Regulation, but high frictions exist with the retailers that are exempted of that network because off this legal construct.

As soon as the car has left the direct influence of the OEM, *Usage and Maintenance* led to new relations to be handled:

- OEM, importer and retailer with the garages
- End-customer, insurance company, garage and appraiser
- Telematics-provider, OEM, end-customer

Here, one might see the telematics-provider on the one hand as the complementor supporting the sale of the car, but on the other building up his own expectations on type and value of the data collected.

At the end of the lifecycle, or at a major disruption point when *Return, Sale or Recycling* come into play, again other relations impose a potential for frictions or even conflicts:

- Between leasing companies
- Retailer and end-customer (e.g. in cases of warranty)
- End-customer, OEM and point of return

Conflicts or diverging expectations and therefore frictions might arise through the obligation to produce the car in a recyclable fashion and the return of the car at the end of its lifecycle.

This not-comprehensive list shows how many interfaces exist that constantly need to be optimised so that those potential frictions do not impose major declines in productivity.

Enhanced Competitiveness

One approach to react to growing market pressure is to extend the collaboration between participants of a common value net. The collaboration between companies and their suppliers, their customers and their complementors is straightforward. By focusing on their key competencies and knowing their place in the value net, the partners will all contribute to the common target of capturing value by providing value to the customer.

However, as soon as substitutors of a common value net, i.e. competition comes into play the roles and objectives in this network become unclear. The many-to-many relations lead to a complex network where the participants might suffer from a lack of transparency and are in fear of single participants treating their self-interest in opposition to the network.

These threats to a successful collaboration between competitors permit frictions in the network and reduce productivity and effectiveness. A mitigation strategy has multi dimensions with two being described in further detail: Introducing a neutral entity and using supporting technology.

A Neutral Entity

In a network of collaboration it could take a long lead time for productive work, if the competing participants developed all relevant standards for a seamless co-operation by themselves. Here, a neutral entity that enables such standards could provide enhanced value by reducing this lead time.

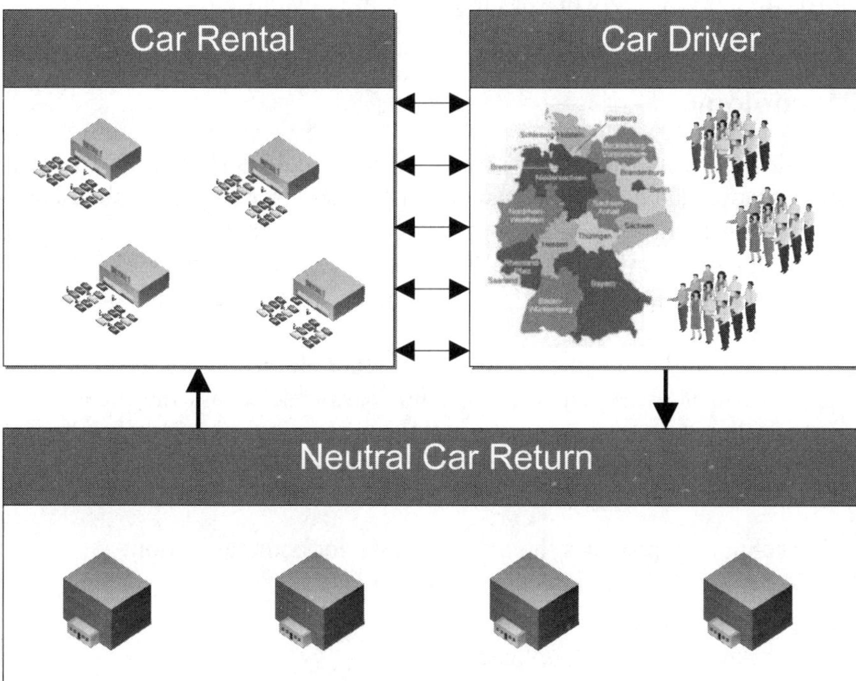

Fig. 1. A neutral entity as the interface to the driver at the end of the rental period can provide both value to the customer as well as enhance productivity for the car rentals.

The neutral entity would need to be accepted by all participants and should not try to create a direct value by itself.

In a different scenario, a neutral entity would create direct value, but would not interfere with the collaborative network. It would bundle all activities of the network partners in a non-core task to enhance productivity, end-customer satisfaction or effectively use scarce resources. In the example (shown in figure 1) the return of rental cars would be handled by a neutral (neutral to the business case of the rental companies and to the end-customer) entity that take the returned cars from the driver, check the condition of the car and prepare it for delivery to the rental company. This entity creates value for the customer, as it provides a single-point of return at locations of scare resources for space, such as at airports close to the gates, does not interfere with the value proposition of the rental companies, as they do not see the car return as a key differentiator, and is paid for by the rental companies.

While more scenarios are possible, one has to be aware, that such an entity does not come without cost and the business case of introducing one into a collaborative network has to be carefully evaluated.

Technology

While partners and other entities within a collaborative network can solve organizational and governance issues, many frictions have their basis in the intra-network, e.g. during the information exchange, at the interfaces and during the communication. The effort to establish networks despite these frictions might be too high to achieve a positive return on the investments.

However, we have achieved a state of technological maturity, where IT support for a network can drive the business case across the break even point and beyond. As a starting point for further investigation and discussion, we have proposed four hypotheses on how technology can fulfill that role:

1. Technology provides the *infrastructure* for communication:
   - It enables automated communication:
   - Machine-to-Machine
   - Application-to-Application
   - It can augment communication
   - Person-to-Person (via videoconferencing, chat, email, …)
   - Add information (via augmenting reality based on digitised knowledge)

- It provides instant communication
- Anywhere, any time, with any device (e.g. wireless)

2. Technology can reduce rework by enabling *shared data* and information:
   - By deploying knowledge management tools
   - Through virtual / interconnected data management systems

3. Technology crosses boundaries by *integrating* distributed environments:
   - It enables community building through teamrooms etc.

4. Technology enables trust by providing *transparency*:
   - E.g. for demand planning and forecasting

It is clear, that the fulfilment of these promises is not without challenges. Each of the hypotheses comes with a task that needs to be handled:

1. Master *complexity* of the infrastructure
2. Solve the *government* of shared data
3. Keep *security* tight
4. Mitigate the *fear* of transparency

However, assuming the feasibility of these challenges, we can postulate, that it is neither the demand itself that hinders the widespread and seamless co-operation between competitors nor it is the infrastructure for the network. It is much more the missing strategy within the business domains that needs to be adapted to include this valid option of responding to market pressure (see Fig.2).

Results of the Discussion Forum

Neutral Entity

The theses sketched in this paper were the basis for a discussion forum at the 2nd e.e.u.c. 2002 and let to a controversial debate on the necessity of a neutral entity. Several participants of the discussion backed their opinion, that such a neutral entity or mediator is obsolete, by their own positive experience of OEM-supplier collaboration without such a third party.

It became clear, that the role and value of such a third party needs to be carefully evaluated and only taken into the network, when otherwise the whole concept might break or the value generated for the customer is overwhelming in relation to the effort.

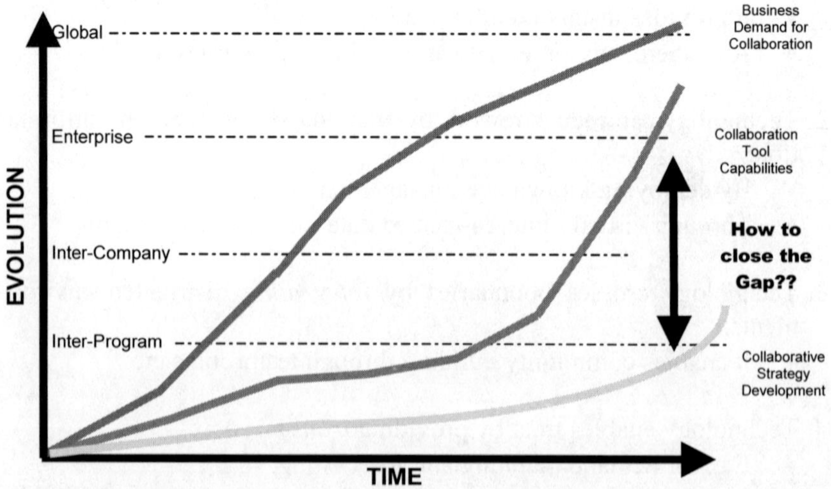

Fig.2: With demand and capabilities in place, the collaborative strategy needs to catch up (Source: IBM Consulting Group Analysis).

Especially in cultures, where trust and unwritten rules enable an effective networking, any additional player would probably not provide additional value, but rather raise cost and disturb the network.

Technology

As the four hypotheses of technology usage were presented in a way to stimulate discussion and objections, the basic understanding was that it cannot only be technology that solves the issues of collaboration. The organizations and people of the collaborating parties are still the major factors for success and failure. To comprehensively describe the aspects of collaboration, the intercultural relationships need to be closely observed and integrated into a collaboration strategy.

Focus Groups

During the discussion the plenum discussed the fact, that there are different approaches and solutions, depending on whether one focuses on local networks of small and medium businesses or on relations between multinational companies, their peers and suppliers on a global basis. As the cultural aspects, roles of mediators and technology might fluctuate widely; currently one needs to describe those focus groups differently.

Core Competences

Recognition, Development and Successful Implementation

Network of Automotive Excellence

Reiner Speiser, Hans H Jung

Nexolab GmbH

1 The Pressing Need for the Automotive Industry to Act

Although the term core competences was coined by C.K. Prahalad and Gary Hamel at the beginning of the 90s in the sense of making economic business-relevant decisions, the subject area became somewhat of a background to the prevalent theme of success factors. Even today, the automotive industry does not seem to have found a suitable approach to this issue.

- Chairmen of automotive companies continue to swear by shareholder value as a central competence for what, in their view, is the key core competence – access to international capital markets.
- Sales & marketing heads describe customer loyalty as the critical success factor and are convinced that they simply will not be able to sell and market automobiles in the future without the core competence called Customer Relationship Management.
- Designers and production directors are clear on one issue: that the core competence Supplier Relationship Management must be implemented across the board in order to deal with future challenges effectively.

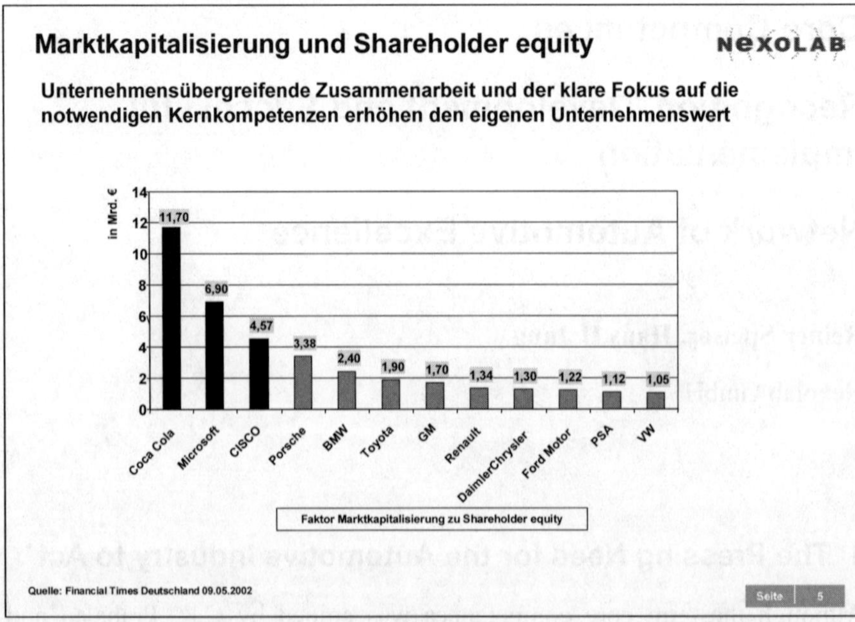

Fig.1: Market capitalisation and shareholder equity (market to book ratio).

Company-internal competition regarding the various areas of competence leads to the implementation of sub-optimum solutions which refer only to divisions within companies and, as far as the automotive industry is concerned, do not go sufficiently beyond the boundaries of the typical company's internal functions.

Companies such as Coca-Cola have been much better than automotive companies at optimising their core competences within value-building networks. Companies such as CISCO have used such a network strategy to become market leaders within a constantly-changing market environment within record-breaking time. Besides applying and implementing standards for activities, CISCO focused on a clear business model, with innovation and engineering competence lying at its heart. All other skills needed have been developed by and together with partners. Using a global customer relationship management strategy and working with key production partners who are responsible for 80% of CISCO's production, CISCO has become a global supplier of network components with 2001 sales in excess of $ 22 billion. In addition to this focus on core activities, CISCO has pushed through 90 acquisitions within 9 years – these have been done to improve market leadership in the area of innovation and engineering.

If we look at successful companies such as CISCO, Coca-Cola or Microsoft, we soon realise that no automotive manufacturer comes close to the market-to-book ratio of these companies. Whether Toyota or Porsche, volume or premium manufacturer, even successful companies such as Toyota cannot compete in this area (see Fig.1)

Even with direct comparisons, no automotive manufacturer reaches a Best Possible Player position.

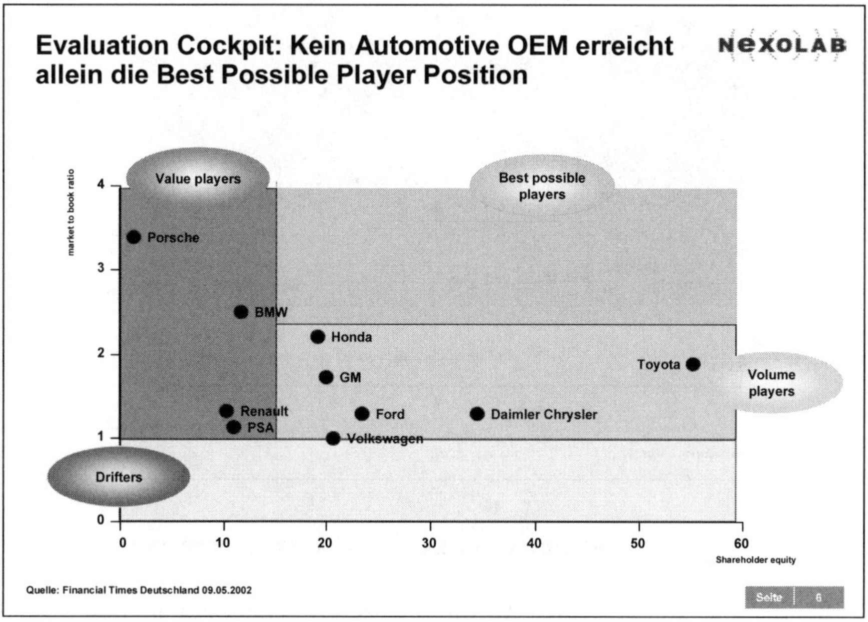

Fig.2: No OEM achieves the Best Possible Player position.

When we consider global suppliers in the automotive industry, a similar picture appears. No single company considered achieves pole position according to financial analysts.

This difference raises the question as to what makes companies such as Coca-Cola and CISCO more successful than automotive companies such as Porsche, BMW or Toyota?

These companies demonstrate considerable differences in their strategic focus in comparison with automotive companies

- Focus on a maximum of 1 to 2 core competences
- Standardisation of processes

- Clear focus on customers and partners
- Transparent Branding
- Compete and Collaborate Strategies within all competence areas.

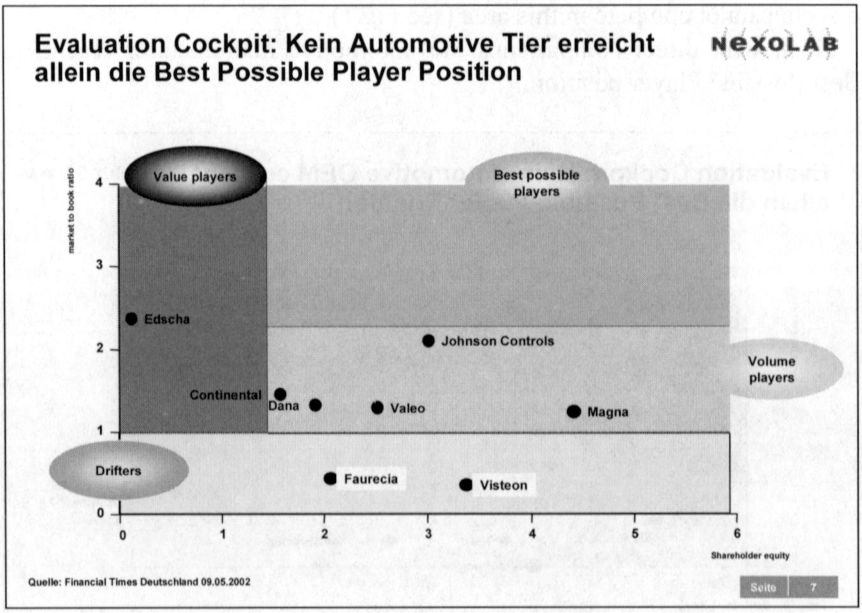

Fig.3: No automotive Tier achieves the Best Possible Player position.

In spite of massive change within the sector, neither automotive manufacturers nor their suppliers have shown or are showing willingness to implement radical change.

2 The Future of the Automotive Industry Lies in Cross-company Collaboration

For many years there have been cooperative models within the automotive industry. But these remain stilted because of a culture of mistrust, pursuance of individual interests and the willingness to see only the borders of one's own company. These factors hinder the ability to profit from synergies across company borders.

According to Payne and Rapp however, one of the critical success factors in current management is the ability to predict market developments and their effects on a company and its complex value chain within a value-building network. Especially in the automotive sector, it is necessary to challenge current management practices in order to achieve potential savings. Also, the following demands mean that automotive companies must identify their core competences and optimise these in order to achieve Best-Possible-Player status:

- increasing numbers of new product launches (see Fig.4)
- ever-shortening product life cycles (see Fig.5)
- customer demands for tailored products and solutions (see Fig.6)
- expansion of traditional business models to include downstream business (see Fig.7)

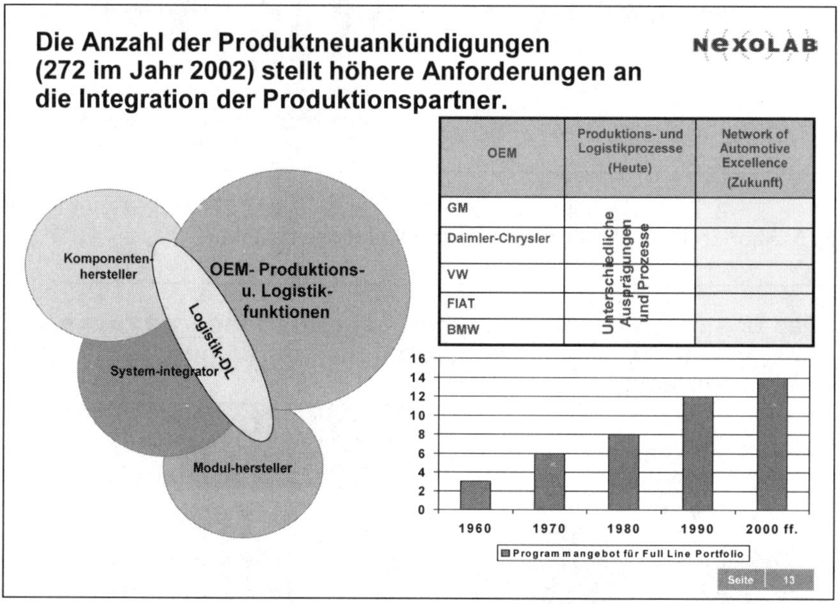

Fig.4: New product announcements demand higher integration capabilities

Both from the viewpoint of original equipment manufacturers (OEMs) and their suppliers, it is clear that missing production and logistics standards are leading to exponential wastage in terms of storage of materials which are contributing to the detriment of adding value – and thus making profits - for all partners involved.

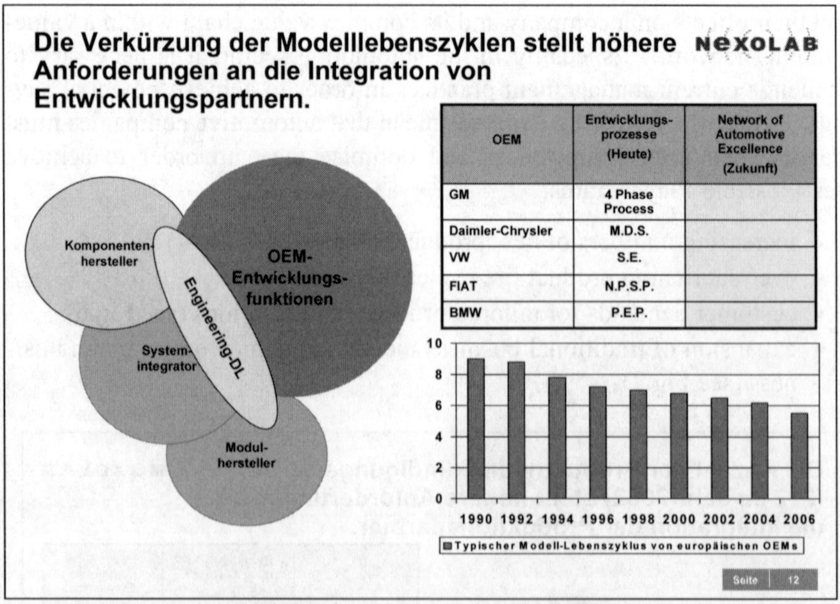

Fig. 5: Shorter life cycles demand higher integration capabilities

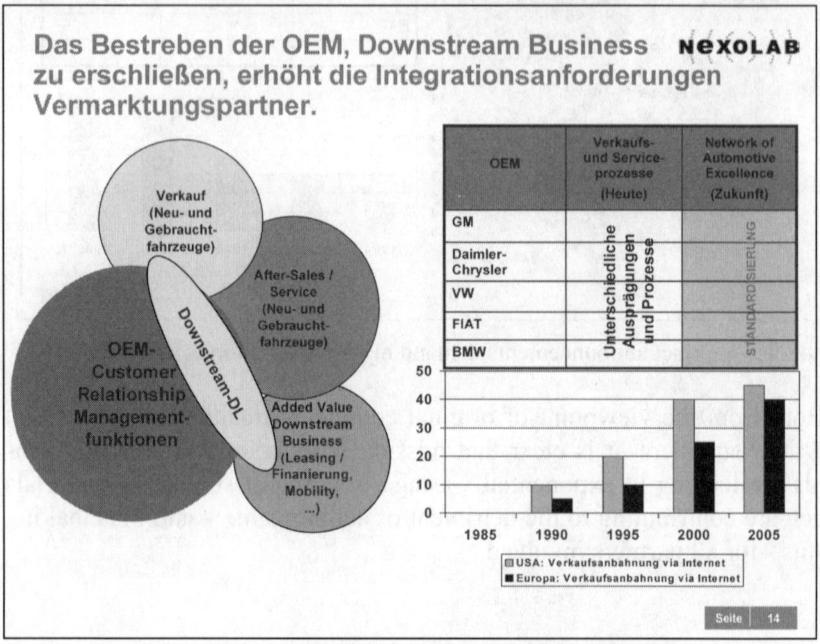

Fig.6: Downstream business demands higher integration capabilities

The requirement for shorter development times as a result of market and customer need mean that product development cycles have to be continually optimised. New development standards are needed as a result of the increasing amount of hardware and software in vehicles (e.g. motor management systems, steering assistance) as well as connections to information and communication networks (e.g. telematic services) – existing processes and systems cannot cope with the pace of these developments. The need to assess investments and the ability to react to market changes within the product development cycle also means that more flexibility is needed.

Within the context of automotive purchase and usage, it is clear that many downstream processes/concepts such as customer-individualised mobility and in-car internet need to be addressed.

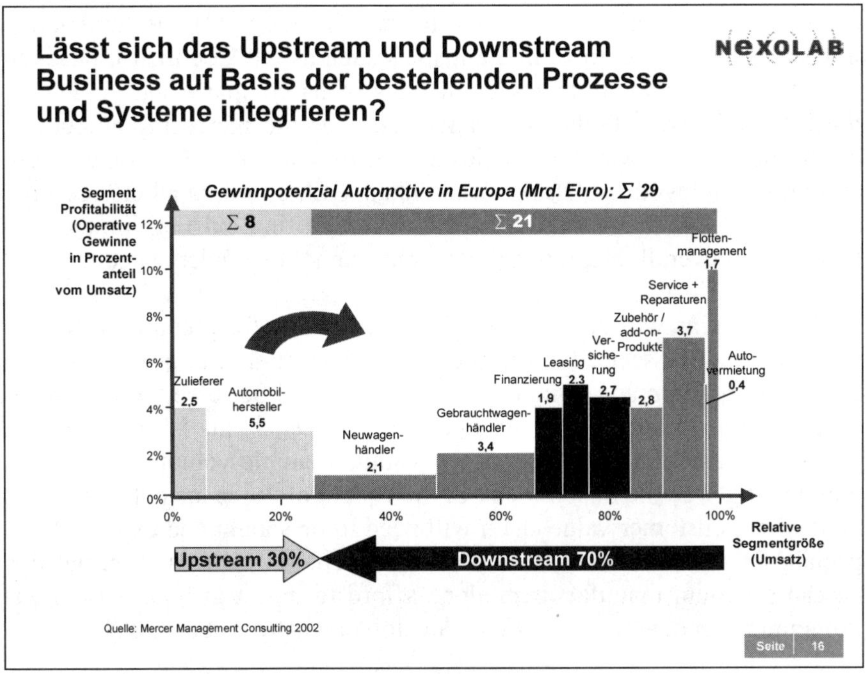

Fig.7: Possible integration of upstream and downstream business?

In order to be able to successfully make money in these areas, it is necessary to build a sustainable and thus profitable relationship to customers, together with content and other partners. Also, the integration of sales activities, distribution and production will be necessary in order to offer complete solutions efficiently and effectively.

When analysing the status quo, it is clear that automotive companies and their suppliers are growing ever closer together in the areas of innovation and engineering and production & logistics. In these areas, strategies for competition and collaboration are in part available, but these are most often dovetailing with the specific needs of the OEM. Every supplier needs to meet the individual requirements of the OEM, or he will not get the work. Whilst product life cycles and units ordered are constantly being reduced, margins are similarly being eroded and development costs being passed to suppliers.

Optimising processes beyond the boundaries of any one individual company is presently rudimentary to say the least. If the automotive industry wishes to learn from companies such as CISCO, then it needs to reflect collaborative strategies in all parts of its company and implement these accordingly. Technical changes such as the increasing amount of electronics and software in vehicles need different competences than those currently employed by OEMs. Individual OEMs and suppliers are only able to work on solutions in a haphazard way using existing limited systems and an overall integrative philosophy cannot be implemented.

High margins will only continue to be available for premium offerings in the automotive sector for a limited time. In ever-shorter product life cycles both premium and volume manufacturers will be forced to earn revenues – competition will get stronger, not weaker. This is no mean feat when we consider that number of units sold or simple volume calculations will no longer be the raison d`etre of the equation. Revenue potential in all parts of the customer value chain will need to be sought and exploited, especially in so-called downstream business. The only question on our lips is: can premium manufacturers alone afford to invest in both new model programmes and, at the same time, develop downstream business?

We come to the following conclusion:

The automotive industry needs an efficient and effective form of collaborative model for future profit generation.

3 Content of Collaborative Cooperation in the Automotive Sector

Die 3. Transformation der Automobilindustrie - Basis für das Network of Automotive Excellence

NEXOLAB

Trans-formations-schritte	Sachlogik und Ausrichtungs-struktur	Interessens-/ Zielsetzungs-struktur	Entscheidungs-/ Verantwortungs-struktur	Verhaltens-/ Führungsstruktur
1. Transformation Verfahrensorientierung innerhalb des OEMs (von der Handwerks- zur Massenproduktion)	Stückzahl und Automatisierung	Optimierung von Teilfunktionen	Hierarchisch	Steuerung der einzelnen, tayloristischen Arbeitsschritte
2. Transformation Produktionsoptimierung aus der Perspektive des OEMs (von der Massen- zur schlanken Produktion)	Prozessorientierung und Informatisierung	Schnittstellen-synchronisation	Delegation soweit notwendig	Steuerung der bereichsspezifischen Funktionen
3. Transformation Simultane Kunden-, Produktions- und Innovationsorientierung über alle Stufen des Wertschöpfungsnetzwerkes (von Built to Stock zum Built to Customer)	Bündelung der Kernkompetenzen und Standardisierung der Kernprozesse	Competition & Collaboration	Selbststeuernde Einheiten auf Basis von Regelkreismodellen	Steuerung der bereichs- und unternehmensüber-greifenden Zusammenarbeit

Fig.8: The automobile industry faces its third transformation

The strategic direction of the automotive industry and its previous attempts at transformation also raise other questions:

- 1st Transformation: Ford achieved economies of scale using a new production philosophy.
- 2nd transformation: Toyota optimised this philosophy within a closed network (with suppliers) with a focus on flexibility and quality.

In the third transformation of the automotive industry, the entire value network of the automotive industry is concerned. We can no longer effectively predict as an industry where future customer demand will lie in terms of mobility, experience, vehicle functionalities, etc. and this leads to the fact that all OEM-specific development, production and marketing concepts will lead to despair. This is shown by the fact that the majority of OEMs even on a global scale can continue to secure revenue by adopting a process of endless discounts to buy customer loyalty. Further reductions in demand for cars result in ever-increasing efforts at concentration within the industry and not to improved circumstances.

Until today, the automotive manufacturer is in the forefront of the focus of the supplier, some might say at the centre of his world. End customers and their needs and desires are only the focus of relatively few automotive suppliers.

The final customer will affect directly all changes within the industry:

- Market proximity and customer retention.
- Mass customisation and order-to-delivery concepts.
- Shortening development and manufacturing cycles
- Increasing product complexity
- Collaborative models

Simply attacking time, cost and quality as many have tried until now will not help automotive companies to proactively shape changing markets and thus anticipate and respond to future customer needs.

Terms such as stable relationships and trust will increase in importance. Every innovation must be measurable in terms of price, function and contribution to revenue; otherwise it will be scrapped.

Stable partnerships with customers mean having stable partnerships with suppliers. This means that suppliers must know and be able to anticipate the needs of end customers in order to be able to secure profitable business with their direct customer, the OEM. Simple cost-saving by either partner is not the non-plus-ultra, but building a collaborative partnership based on trust.

4 Describing the Network of Excellence

Networks are an intelligent form of collaboration for companies which when created in an open and transparent way, can hardly be beaten in terms of effectiveness and efficiency. These "open" networks cannot serve the needs of one individual company. For Japanese companies, this means that closed networks or keiretsu will no longer provide the basis for success they once did.

Within networks, changing needs and capacity requirements can be reflected:

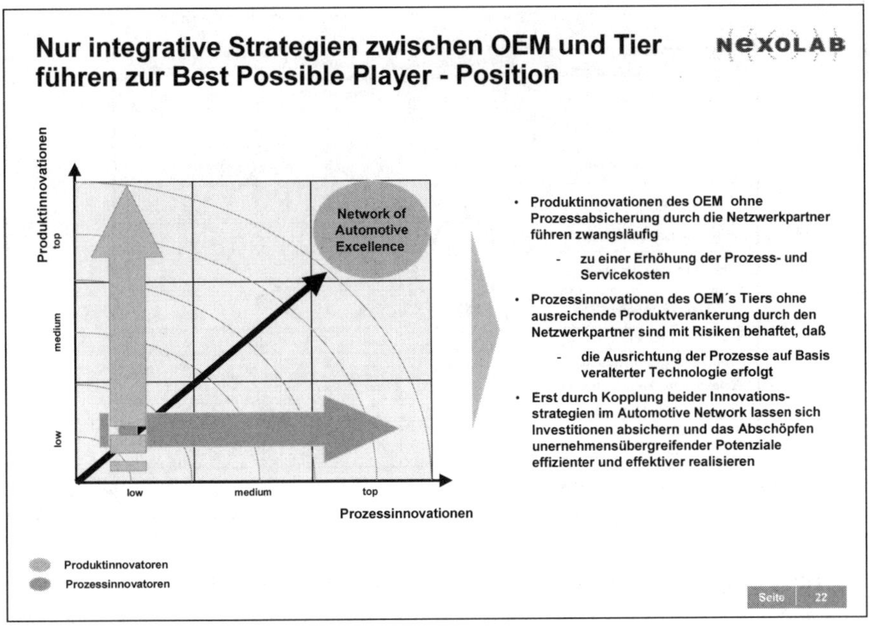

Fig.9: Integration strategies – Network of Automotive Excellence

Existing product or process innovation strategies usually have one aim. BMW and DaimlerChrysler for example focus on product innovation, PSA and Ford on process. Following both strategies at the same time is extremely unusual, but Toyota tries this. Achieving an overlapping strategy focussed on both product and process is not something of which either OEMs or their suppliers seem capable at present.

Synchronising strategies can occur with best effect in an open Network of Excellence, which permits the combination and usage of core competences on a flexible basis as and when needed. The prerequisite for success is, however, being able to be flexible with the entire automotive value chain and this is not easy.

Kooperationsmatrix im Network of Automotive of Excellence aus Sicht eines OEM [1] NEXOLAB

Funktionen / Kooperationspartner	Innovations- und Entwicklungs-Management	Produktions- und Logistik-Management	Markt- und Kunden-Management
Wettbewerber (weitere OEMS)	X	X	-
Systemintegratoren	X	X	-
Modulhersteller	X	X	-
Komponentenhersteller	X	X	-
Engineeringsdienstleister	X	-	-
Logistikdienstleister	X	X	X
Retail	-	X	X
Forschungseinrichtungen	X	X	X
Telekommunikationsanbieter	X	-	X
Energieversorger	X	-	X
Finanzdienstleister	-	-	X
Organisationen u. Verbände	X	X	X

[1] siehe Vortrag Prof. Dr. Milberg – Automobilforum Graz 2001

Seite 23

Fig.10: Co-operation within the Network of Automotive Excellence

The content of a Network of Automotive Excellence cannot simply be product, process or competence-driven but has the following issues also:

- Company and network culture
- Business and benefit sharing models
- People and qualification
- Knowledge-Sharing
- Relationships
- Organisational change

5 Results of the Discussion Forum – Network of Automotive Excellence

The challenge described here in overview is seen by managers much as a haze. This simply reflects the fact that clear management strategies are not available:

- Company-specific solutions define the approach of management
- Even where management has new ideas, decisions are not taken on the basis of all potential approaches and available information. The

multiplicity of conventional business models and calculation matrices reduce complexity but simplify potential and thus sometimes lead the way to disaster.

The cause of intransparency within companies can often be found within internal think-tanks which should actually be working on innovative solutions but more often are relegated to more moribund activities. Radical increases in effectiveness and efficiency are available within Networks of Automotive Excellence but the following consequences must be appreciated:

- Cross-company concepts for market proximity and customer retention are needed,
- Network-specific integration of sales and productive process in terms of mass customisation must be followed,
- Common concepts for sourcing and cooperation with suppliers and logistics companies are needed,
- Development and production cycles must be linked to sales and procurement cycles.

Integration within a Network of Automotive Excellence will be the prime contributing factor toward meeting the challenges of the future and successfully, proactively dealing with these. These issues were addressed by the following individuals at the e.e.u.c. 2002, using examples from their own companies:

- Wilhelm Becker, BMW Group
- Dr. Peter Stehle, Freudenberg
- Peter Köpf, ZF Friedrichshafen

In an in-depth discussion with the eighty participants of the workshop, various key themes necessary for building a Network of Automotive Excellence were put forward. These can be resumed in six core theses:

1. The core inputs and the complementary competences of the partner enterprises must be clearly defined
2. There must be an early integration of partner enterprises
3. Partnerships must be orientated towards the long term
4. Partners must be willing to share knowledge on the basis of benefit sharing models
5. Clear goals and rules for collaboration must be established
6. There must be high mutual acceptance among the partners.

The discussion of terms such as trust, acceptance as well as benefit sharing models took up a large share of the discussion.

The participants passed a roadmap for the building of a Network of Automotive Excellence which is depicted in Fig.11.

Roadmap des Network of Automotive Excellence NEXOLAB

Termine / Aktivitäten	e.e.u.c. 2002	Next steps I. Quartal 2003	Next steps II. Quartal 2003	Next steps III./IV. Quartal 2003
Vision	Präsentation der Vision	Konkretisierung der Vision	Verabschiedung der Vision	
Strategie	Diskussion der Zielsetzungen und Inhalte	Konkretisierung der Strategie	Festlegung der Partner (Phase I)	
Kommunikation	Referate und Workshop	Aufbau Kommunikations-plattform	Veröffentlichung eines Readers Network of Automotive Excellence	Breakout Session auf Fachtagungen und Erfahrungsaustausch
Organisationsform		Aufbau eines gemeinsamen Programm-ausschusses	Konstitution der Organisationsform Network of Automotive Excellence	Ausbau themenspezifischer Communities
Handlungsfelder	Diskussion der Handlungsfelder im Forum	Einreichung von Projektvorschlägen / Initiativen / Best Practice Cases	Implementierung des übergreifenden Programm Management	Durchführung des übergreifenden Programm Management
Umsetzung		Identifikation von Pilotprojekten	Bewertung von Pilotprojekten	Start von Pilotprojekten

Seite 31

Fig.11: Roadmap of Network of Automotive ExcellenceThis roadmap will be elaborated in the next few months by ewf europe and Nexolab GmbH, and will be discussed with the potential partners of the Network of Automotive Excellence. The goal is to obtain first experiences with the approach of Network of Automotive Excellence within the context of one or several pilot projects, and to put these at the disposal of a broader public.

References

Prahalad, C.K., Hamel, G. (1990): The Core Competence of the Corporation. Harvard Business review article

30 Per Cent More Efficiency in Product Development

Fiction or Hard Fact?

Rainer Kurek

Automotive Management Consulting

Dynamic changes have been making their mark on the motor industry for some years. Manufacturer's demands on the development partner are as drastic as they are clear:

Structural performance adjustments and organisational rearrangement alone are not enough. The 6th discussion forum of the 2nd European Engineering User Conference shows the solutions currently under review by experts.

A progressive saturation in European, North American and Japanese markets is exerting enormous pressure on all motor vehicle manufacturers. As a result, there is a drive to seek out and capture niche markets. A continually increasing variety of models and versions is driving the industry towards a multi-project landscape which it can no longer control using existing structures and processes. The available resources of time and money are not sufficient to develop internationally competitive products. Until today, the key focus has been on improved productivity strategies and attractive innovations. Now the shift is towards increased efficiency in the development process. *»30% more efficiency!«* is the current mission statement of the leading European motor vehicle manufacturers in the field of vehicle development.

Programme Management Centres as a base of resource

As an initial step many manufacturers have set up their own Programme Management Centres. They go by various names, e.g. SE or design house, but all serve the same purpose. To compensate for their own lack of internal resources, the motor vehicle manufacturers (also called OEM; Original Equipment Manufacturer), integrate qualified external development partners in product development, and occasion them to be physically or at least virtually on-site, to work together in the Programme Management Centres. The ability to deliver pertinent products to the market at the right time (Time to Market) has now turned into a necessity for motor vehicle manufacturers. The Programme Management Centre is mandatory to achieve this objective.

The goal-orientated combination of the core specialties of the OEM and the diverse competencies of its sub-contractors leads to an integrated development process which boosts the product impact in international competition. Using and incorporating the sub-contractors' innovation potential in the development further enhances success. The ability to respond and the focus and quality of decision-making during projects are thereby all enhanced – these significant improvements are the result of close collaboration within the Programme Management Centre.

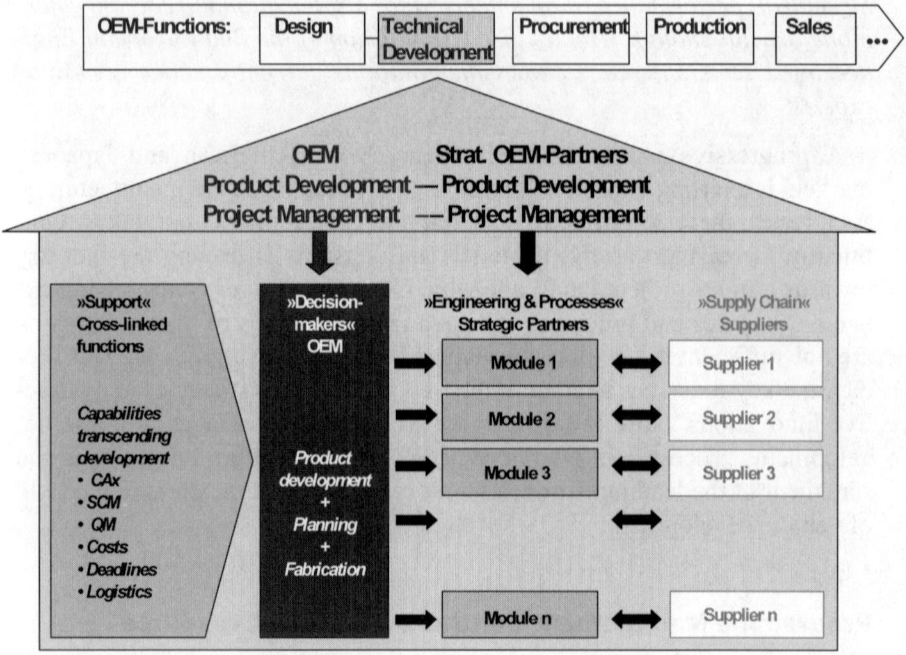

Fig.1: New form of collaboration through integration of strategic partners into the development process

However, that is still not enough for a marked increase in efficiency. In order to understand why, we must consider more closely the concept of a project company (Fig. 1). In the example shown, the strategic OEM partner undertakes the selection of the sub-contractor for all modules of a vehicle. In order to make structures, interdependencies and functions clearer, several modules can also be combined into one system (equipment, chassis, etc.). Unlike before, the strategic OEM partner not only processes but also actively guides the project. The role of the strategic OEM partner can be undertaken by either the development service provider or the qualified supply chain sub-contractor.

Corresponding interface functions (methods and tools) are in place to support them in realising this task. Critical for the success of the project is that, as process leader, the strategic OEM partner uses to the full the development know-how of motor vehicle manufacturer, who is de facto the decision maker. The goal-orientated interlocking of product development and administrative project management processes has largely taken place, laying the foundations for the best possible conditions for a successful transfer of large-scale projects. But are these organisational improvements really enough to significantly enhance efficiency in the development process? Will the use of Programme Management Centres eliminate shortcomings such as inadequate communication, poor change management or ill-defined responsibilities? Certainly not to a great extent!

With process optimisation to the summit

Transferring large-scale development projects to qualified external partners in Programme Management Centres is only part of the solution. The design of the development processes themselves is every bit as important for sustained improvements in efficiency. However organisational matters are also closely linked. One example is timing the linking of the technical competence from tool planning and production and quality control into the process of the product. Nonetheless, linking alone does not guarantee success. Various tools are mandatory to improve the quality of information for everyone involved and ensure a standardised, centrally controlled information source.

Effective "change" management for all departments involved is a further prerequisite for a smooth-running SE project (Simultaneous Engineering). As central process leader, the development service provider is the focal point for information and so must take ownership of the Product Data

Management (Fig.2). To be able to take on this responsibility he must have authority to make decisions such as those relating to a collective PDM system (for the Product Data Management). The planned start of manufacturer can only be guaranteed with professional change management and use of standard software (e.g. DMU = Digital Mock-up).

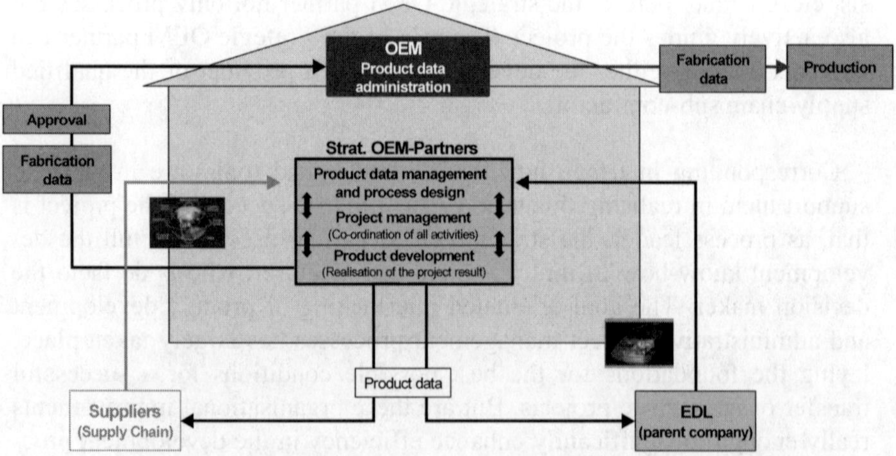

Fig.2: Data and information transfer between development partners (PDM)

It is important for the OEM to nominate all sub-contractors quickly to avoid fundamental changes and to reduce the risks in the active development. And at this point we find ourselves faced with an inconsistency that can only be resolved by clearly defining and assigning tasks, authorities and responsibilities. Since the nominated sub-contractors often assign development requirements to the development service provider, everyone involved in the project must accept the authority of the Programme Management Centre, as defined centrally by the OEM partner.

Where the contracted development service provider is also the strategic OEM partner and acts as process leader, there is potential for conflict between the development service provider and the other sub-contractors nominated and paid by the OEM. The nominated sub-contractor does not always accept the process leadership role of the development service provider, with whom there is no business relationship, and this situation frequently causes friction. The benefits of synergy from this concept can only really be enjoyed if the motor vehicle manufacturer delegates the relevant authorities to its strategic partner, or to the development service provider, if appropriate, who then assumes the role of process leader.

Discussion forum, Brussels

The aim of this discussion forum, which took place within the context of the 2nd European Engineering User Conference in Brussels, was to sit, everybody involved in the development process, around a table to discuss the situation described above and suggest intelligent solutions for future collaboration. In order to keep objectives and goals firmly in mind throughout the discussion forum, and above all to extract the maximum contribution from participants, a study - *»30% More Efficiency in Product Development«,* which collected the practical experiences of executives in the motor industry, was carried out before the congress. The *»Eff 30«* Management Model (Fig.3) is a collation of the findings collected during the preliminary study.

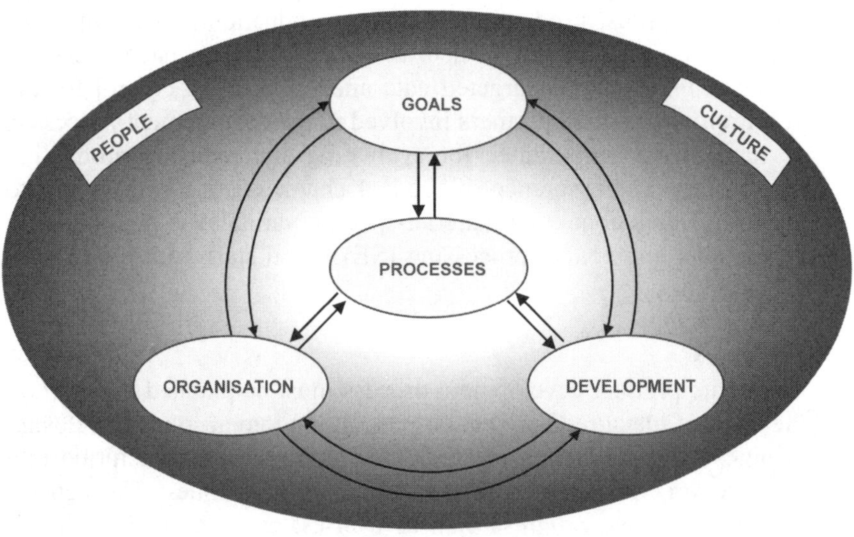

Fig.3: Eff 30 – system-orientated management approach for increasing efficiency within product development

The focus during the development of this model was to create an integral summary, as near practice-orientated as possible, of the actual results generated with the help of the study. The model primarily takes into account the new role of the human factor in individual productivity as well as clear requirements with regard to target agreements, product production, product form and organisational development. The core statements can be summarised as follows:

1. *Goals*
   Before the start of each project, the requirements and project targets should be agreed and specified as exactly as possible between all parties involved in the development process. Banal as that sounds, experience shows that specifications and contractual agreements with regards to technology, carry- over parts, deadlines and costs are usually vaguely defined. This leads to friction in the development process.

2. *Development*
   Defined CAx technologies should be applied consistently and universally throughout the entire development process to ensure the availability of product information for all participants at all times. An integrated development process (i.e. early linking of prototype production, trials, production planning, production and quality system) can be realised only when all named faculties have access to common information. Targeted data and information control for the synchronisation of all partners involved in the development process is of fundamental importance for evaluating and realising production capabilities. The assurance of product characteristics with regard to geometric, functional and material-specific variables is possible only via parallel component processing (SE) by all staff with product responsibilities.

3. *Processes*
   The entire product development process must be planned in detail on the basis of the agreed project targets, taking into account all relevant changes and pro-actively steered. This requires clear structuring into defined work packages with corresponding milestones. The generation of a detailed progress plan as a »Best case« scenario helps to carry out systematic risk analyses with regard to methods, deadlines and costs during the project. Systematic change management on the basis of early indicators is necessary to facilitate pro-active project continuity and reduce friction losses.

4. *Organisation*
   Although a Programme Management Centre can facilitate significant improvements in cooperation, the tasks, authorities and responsibili-

ties must be specified clearly and unambiguously for all development partners within a flexible and transparent project structure. Function diagrams help to define rules which simplify the collaboration of the different participants.

5. *Human factor*
   The employee is the essential factor for successful inter-company co-operation and communication. A joint and learnable culture must be developed for the employee. The new, changed basic conditions for the product development process require competent, motivated team players who will pursue common values and goals throughout the project collaboration. Experiences must be continuously gathered in a "knowledge bank" according to clearly defined learning goals and processed and made available in a suitable form for future projects.

Fig.4 below summaries the essential findings of the preliminary study:

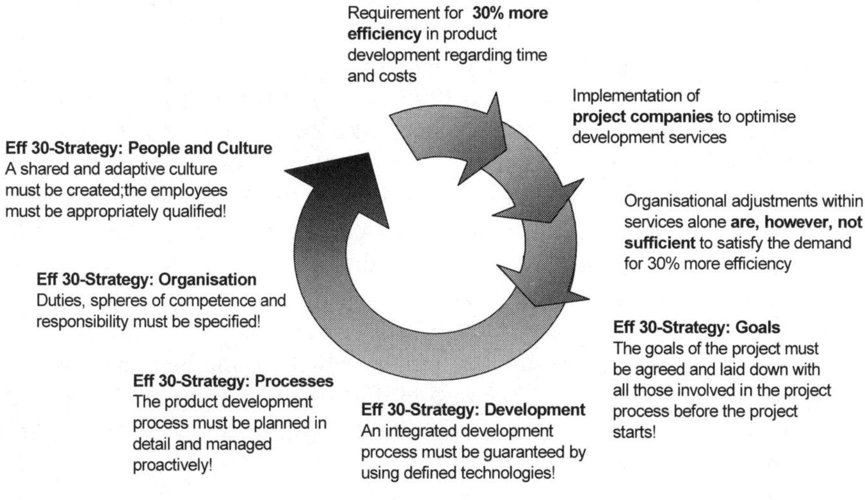

Fig.4: Results of a preliminary study on the subject »30% More Efficiency in Product Development – fiction or hard fact?«

We were lucky enough to have present at the discussion forum Dr Weiss (AGENSIS company consultancy), Mr Balk (former VW commercial vehicle development manager), Mr Dokter (EDAG Board of Management) and Mr Wagner (Automotive Management Consulting), in other words, some extremely competent industry partners for this working group. The discussion forum was planned in such a way as to be able to address the topic from different angles. Dr. Weiss produced current market investigations to confirm that it is possible to considerably increase efficiency in product development by optimising processes and project management. Mr Balk outlined what the OEM developer expected of its development partners in order to achieve 30% increased efficiency. Standardisation of quotations and improvements of the design specification on all sides were treated as possibilities for optimisation as was the necessity for the development of clear areas of sub-contractor authority. Mr Dokter for his part outlined what the engineering service provider required of the OEM. The OEM purchasing process, the delegation of responsibilities and long-term strategic cooperation between the OEM and the ESP were all picked out as central themes and proposed as areas with potential for increased efficiency. The focus of the talk given by Mr Wagner was the productivity of knowledge processing or the new role of the human factor. Effective strategies for more productivity, unlocking potential, promotion of cooperation and securing the knowledge obtained from the experience of employees were central components of his presentations. Overall, a series of interesting aspects of strategic partnerships as well as collaborative networks was gathered which requires thorough intensification. Increasing efficiency in product development has become a major challenge for the motor industry. It is essential for the successful placement of products in the worldwide competition. Therefore the results of the discussion forum will be turned into a fully scientific study which should be published in mid 2003. The combined results of the preliminary study and the discussion forum can be summarised as follows:

1. The purchase order from the motor vehicle manufacturer must be concise and detailed (clear design specification standards). Ground rules for basic changes to design and technology must be agreed before the start of the project.
2. Tasks, authorities and responsibilities assigned to the motor vehicle manufacturer, development service provider and sub-contractors must be absolutely clear when assigning the project and should be confirmed in writing (chain of command).
3. The change management system (methodical linking of methods, deadlines and costs) should be centrally led and pro-actively controlled. Clear provision of data at pre-agreed times (synchronisation)

serves as a common reference for design, accounting, planning and all other departments involved in the development process.

4. After assigning the project, the domains of control and management should be defined within the scope of the project organisation, giving clear direction concerning decision-making powers.
5. The technical structure of the project must be crystal clear for everyone involved, so they can understand system dependencies in the course of the project (effects on development budgets and deadlines).
6. Before starting the project, the quality of data as well as the type and frequency of information transfer should be agreed and definite points of contact should be identified.
7. The strategic OEM partner must be qualified to manage the design, development and control of a supply chain, in order to be able to implement the right structures and processes in the development of the product, keeping the objectives and results firmly in mind. Figure 5 shows the essential components in successful supply chain management. For the process leader, being able to adopt a proactive leadership role in the development process means above all having the competency of integration at his disposal, in order to be able to use the benefits of synergy in collaboration with the sub-contractors.
8. Product development and project management must be interlocked as defined core processes, to enable the development of inter-company collaboration.
9. The new role of the employee must be recognised, and corresponding qualifications generated.
10. Both parties (OEM and service provider) must agree on the qualification of the employees concerned.

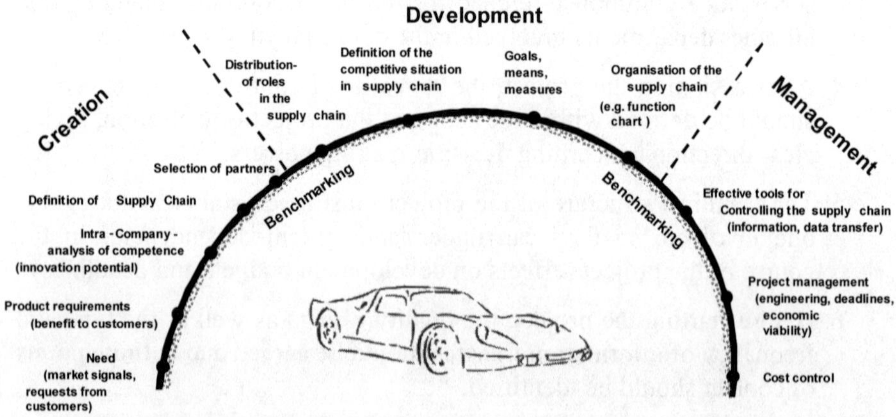

Source: cf. Kurek, R. / Schindler, S.: *"Gewinner von morgen handeln heute"*, DVA 2002.

Fig.5: Creation, development and management of a supply chain

Productivity of "Knowledge" Work or the New Role of Men at Work

Reinhard Wagner

Automotive Management Consulting

In September a press report published by Gallup drew attention to bad management and a resulting lack of commitment in employees. Accordingly, 85% of German employees do not feel any real obligation towards their work, they lack commitment, 16% are even actively uncommitted, i.e. they are not productive anymore or have even given up trying completely. Reasons for the lacking commitment in employees, as stated in the report, are as follows: „People do not know what is expected of them, their superiors are not interested in them as human beings, they often work in a position which is unsuitable for them and their opinions and views do not seem to be important."

These statements are particularly alarming, as people play an ever-increasing role in a society based on information and knowledge and they make a tremendous contribution to the productivity of companies. This has not always been the case. In an agricultural society, productivity was primarily dependent on the experiences of living with nature. Starting with the industrial revolution, human work was transferred into the workshops of industrial plants and by means of an ever-increasing degree of mechanization, productivity has continued to achieve new records until today. Due to modern information and communication technology, it is brainwork, i.e. the creation, processing and evaluation of knowledge, which gains in significance. The productivity of "knowledge work" is directly connected with people, their motivation, qualifications and opportunities to develop their skills.

To increase productivity a new understanding of leadership is necessary. It is the conduct of leadership that is decisive in proving, whether the employee will operate effectively in his surroundings or not.

Allowing potential to develop

It is the primary task of leadership to develop the potential of people to their best possible performance. In this context three important aspects of human performance must be considered: The willingness to perform, the opportunity to perform and the ability to perform, put simply "wanting", being "allowed" and being "able" to perform. Numerous offers of seminars dealing with personal motivation show that many people in leadership positions distrust the motivation of their staff and tend to rely on external motivation. The natural motivation of people can be destroyed if too many incentives are offered. The primary task of leadership is to build up a connection between the task requirements and the employee's expectations by careful selection of personnel and showing respect towards their aims. Personal development must be possible and also room for the employee to show initiative must be created in order to promote skills. Performance skills require a certain degree of commitment on behalf of the employee and also well-oriented support from the company in providing sound training and continuous development of the employee's personal qualifications. In this context various aspects are to be promoted, such as aspects concerning a specific field of knowledge, defined methods, social and personal aspects. Individual training and coaching or specific career models support the development of an employee according to his needs.

Promoting cooperation

Modern information and communication technology offers unlimited possibilities of doing business with partners all over the world. Therefore, companies must find their way out of the closed, integrated structures of the past and become cooperative network organisations communicating permanently with various partners on different levels of a supply chain. As a consequence, companies have to establish open structures and develop their capacities for cooperation and communication. A prerequisite of successful cooperation beyond a company's limits is to agree on common values and rules, sometimes even standardize symbols.

As tasks become increasingly complex, efficient cooperation is required, especially between team members. The primary task of leadership is to provide well-defined goals and promote commitment. Tasks, competences and responsibilities have to be clear within the overlapping structures of business. Binding rules within a team ensure smooth cooperation. The selection of qualified team members and the active formation and development of the team are further tasks of leadership. Furthermore, it is important to provide transparency within a project. For that purpose the supply of necessary information as well as regular feedback on the current status in a project are needed so that foresighted leadership will be possible even in difficult situations.

Learning from experience to create new perspectives

In our knowledge-oriented society, manual work is increasingly replaced by brainwork. Knowledge and creativity are the factors of production in a company of the 21st century. The modern media put an abundance of information at our disposal, but at the same time the half-life period of knowledge also rapidly decreases. By investing in databases or other technologies many companies try to retain the flood of information. However it is the human factor, which often fails when applying a technological approach. Organisations act through people. Knowledge is also bound to people. For knowledge is experience which a person has gathered in a certain context. If knowledge is to become effective it is necessary for the employee to become effective within the organisation, i.e. the conditions for organisational learning and innovation must be achieved. Only in this way can we successfully adapt to the changing conditions and increasing pressure of today.

Leadership, then, has to admit personal creativity, and with a high degree of social skills, allow for an intensive exchange of experiences. Important factors for success include orientation towards higher goals, the understanding of dynamic and linked systems, a positive attitude towards learning, the ability to work in a team and tolerance as well as flexible but methodical action.

The slowness of change within organisations very often is an obstacle for flexibility and progress. It is often the employees, caught up in traditional organisational structures and rules, which oppose change and rather seek security in routine and repetition. Leadership should, exactly at this point, start to promote willingness for change amongst participants, by creating an atmosphere of trust, by supporting self-esteem and by increased delegation and participation. In contrast to material matters, sharing does not decrease knowledge; it increases it. It is not the possession of knowledge that is important, only constructive dialogue between the participants and the exchange of knowledge ensure progress in the sense of more productivity and innovation.

Leading instead of managing

The prerequisites to human work have drastically changed from the times of industrial revolution to the knowledge-oriented society of today. In the past diligence, discipline and subordination were demanded; nowadays it is rather autonomy, ability to work in a team and flexibility. These changes are the result not only of objectively modified working conditions and the change of values, but also of the new understanding about the role of men at work. The role of leadership also has to be redefined in such a modified context. Quickness and ability to innovate, flexibility and a close contact to the customers can only be achieved if responsibilities are delegated and the staff is admitted more personal autonomy.

The new role of leadership requires setting aims instead of giving detailed instructions, to trust the staff and regard people as equivalent, i.e. recognize them as entrepreneurs in the enterprise. If the staff is considered to be dependent, that is inevitably what it will become. To stop this vicious circle means to mobilize unknown potential. Consequently leadership means building confidence, forming trust, creating Challenges and promoting development. Peter F. Drucker summarized the purpose of leadership in the 21 century as follows: „People are not managed or administered - the leadership of personnel is the decisive task. The aim is to transform the specific qualities and the knowledge of each person into productivity.“

30 % Higher Efficiency in Product Development

Requirements to the OEM from the Engineering Service Suppliers' Point of View

Karl Dokter

EDAG Engineering + Design AG

The request of higher efficiency in the product development process by the customers of the Engineering Service Suppliers is not new and they are prepared to take it on. In reverse they demand some key requisites from the OEM to achieve this. Only a common approach can secure the target. Some of these success factors are pointed out in the following explanations.

Relative freedom of process

Most suppliers established their own processes and they are continuously improved to survive in the market. They will reach the highest efficiency by using these processes in allocated projects. If they are forced to implement to many new workflows in the established ones their advantages vanish. A detailed clarification has to take place in the preliminary stages of the project.

Integration of resources in low cost countries

The outsourced work volume varies from project to project. But fortunately the projects gained in volume and last for longer periods. At the same time competent resources were build up in some foreign regions like India and Eastern Europe. Further cost advantages and an additional capacity can be activated by allowing the utilisation of these cost efficient resources of the Engineering Service Supplier. The outsourced project must have a certain size to be able to integrate these resources. Otherwise the additional interface reduces the possible positive cost effects.

Integration in the complete Development process

The highest effect is achieved by integrating the maximum number of available competences at the Engineering Service Supplier. This reduces interfaces to the OEM and allows the Engineering Service Supplier to activate established procedures between his competences. By cutting the project in to many single projects at different partners the risk of coordination errors increases and the OEM still has to deal with most of the interfaces without freeing his managing capacities.

Early integration into the product generation process

An early integration of the Engineering Service Supplier into the projects will reduce risks of missing information. The integration already in the concept phase makes sense, because important fundamental decisions, affecting the entire development process, are made in this phase. The Engineering Service Supplier already gains fundamental project know how. Subsequent decisions are based on this early experience and information. Also increases the volume of the required information along with the project duration and it has to be transmitted at a late start of the collaboration. In practice this is not possible caused by the lack of time.

Allocation of the external volume

As external development work is sourced like the supply of the mass production parts an important role is assigned to the purchase procedures. The purchasing process of the OEM should begin at an early stage and should be completed before the beginning of the project because

- contents can be well co-ordinated
- interfaces clarified
- capacities of the Engineering Service Supplier can be planned optimally
- necessary employees of the Engineering Service Supplier can be taken out of a current project smoothly
- the project infrastructure can be optimally prepared

All these factors would cause disruption in the project work if left unsettled before the start.

Fixing of responsibilities

The sharing of responsibility is one of the most important issues in the outsourcing process. Most of the problems in running projects are the result of not finished or even not started responsibility discussions. These clarifications have to be made profoundly and early enough. This excludes time consuming discussions during the project and gives a good working climate.

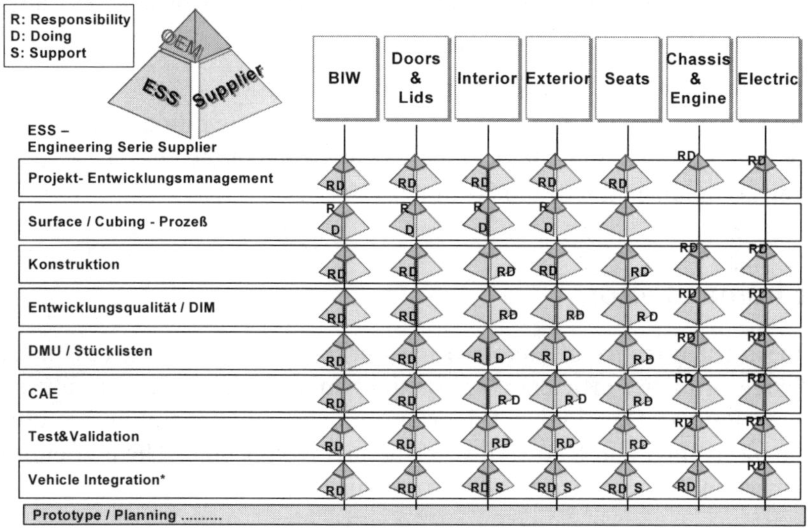

Fig.1: Responsibility

Continuity in partnership

There are many suppliers for a dedicated service available in the market. So shopping around and using different new partners for the same task could be desirable strategy. But there are some hard arguments to look for a long relationship of a competent Engineering Service Supplier as this provides the following advantages for the OEM:

- the process know-how of the OEM stays in an area he has under control
- the knowledge remains with the Engineering Service Supplier and is activated from project to project

- the Engineering Service Supplier already knows the interfaces between himself and the OEM

This will bring efficiency into further projects and speed up the ramping from project to project.

Direct IT connection to the OEM

The need of exchanging data has increased and risk minimisation and reduced lead times demand an effective data exchange. All the involved partners must have immediate access to the correct data. To achieve a fast transfer of the required product and process data with only a few interfaces, the Engineering Service Supplier needs

- direct access to the EDM/PDM system of the OEM
- direct access to the release system of the OEM (condition is a contemporary co-ordination with the release responsible of the OEM)
- open communication of the IT-strategy

The human factor

All these listed technical factor will not bring the expected success in the challenge of higher efficiency unless the human aspects are also taken in account. Product Development depends and still will depend in the future on the motivation and Know How of the involved people. The ewf europe set a mark in locking the human factor to his IPOT strategy.

The most important factors to reveal potentials for efficiency increase are Confidence in the partners and Freedom to shape the path.

Virtual Level Collaboration: SMEs in Aerospace

Stefano Carosio

D'Appolonia SpA, Genova (Italy)

The adventure of space has led to significant scientific advances. New technologies and materials have had to be developed to be able to cope with the extreme conditions or to handle the specific problems faced by space exploration. (High performance optical devices, radiation-hard components, powerful simulation tools, etc.). Many of these new technologies and materials can be adapted for use in more "down-to-earth" applications in a multitude of fields: medicine, transport, mechanical engineering, mining, the textile industry, and many more.

This contribution contains elements of a workshop held at the conference.

The ASLINK (Aero-Space Link) project , of which D'Appolonia as acts as Co-coordinator, has been the first Economic and Technological Intelligence (ETI) project funded by the Innovation and SMEs Programme as part of its specific measures for SMEs. Its aim was to assist SMEs in analysing and exploiting market opportunities in the aeronautics and space sectors and to support them in preparing proposals for funding under the European Commission's Fifth RTD Framework Programme.

The original impetus for the project stemmed from the European Space Agency's (ESA) technology transfer programme which has been running since 1990. A core group of European aerospace technology brokers (Spacelink) saw the need for wider-ranging support, specifically aimed at SMEs, which would look not only at the straight transfer of technology but also at the funding of proposals for the adaptation and re-engineering of key technologies, thus reaching the critical mass to fill the gap with final end-users. A joint collaboration between the European Commission and the ESA was initiated in 1998 and resulted in a series of ongoing initiatives - of which ASLINK was one of the first.

The success of the ASLINK project has surpassed all expectations. In the two years since the project started, 50 proposals (including six Exploratory Awards, one Thematic Network, 4 RTD applications, a pilot action application for Collective Research, and thirty-eight CRAFT Proposals), worth some 63 million Euros and involving over 80 aerospace companies and 180 non-aerospace companies, have been submitted to the European Commission. The evaluation process is still underway, but so far over 50% of the proposals evaluated have been accepted for funding. A promising development concerns bicycles. Although electric bicycles have been on the market since last ten years, they are not much diffused because of the high price when compared to reduced range performances, the high weight with respect to traditional bikes and the short life cycle of the battery; but they could reduce a part of the problem of the pollution in the European cities. To reduce this obstacle there is a need for a lightweight, long range and cheap electric drive system that can be simply retro-fitted to all traditional designs of bicycles. The project objectives are to create a high efficiency electric drive unit using an innovative combination of Li-polymer space battery technology adaptable from the European Space Agency and a novel nutating reduction gearbox, that could be retrofitted to the majority of the estimated 1,5 billion bicycles currently in use in the world and the estimated 70 million new units entering the market each year.

Within ASlink, T4Tech, a Spin-off Company of D'Appolonia was created. T4Tech provides highly qualified services to SMEs concerning the recruitment of funds for RTD activities, in order to strengthen their competitiveness on the market. Besides this, T4Tech provides technological consultancy, acting in close synergy with D'Appolonia, able to involve its 150 engineers in multidisciplinary technological sectors. Within T4Tech activities, the first European Network of Space Incubators will be officially launched on July 17th 2002. Both Mr. Busquin (EU Commissioner for Research) and Mr. Rodotà (Director-General of the European Space Agency) will attend the event, thus representing an important signal of the strong interest of both EC and ESA about ESINET. The ESINET is in fact strongly supported by the European Commission and European Space Agency. The ESINET builds on the new ESA-EC common strategy for space, targeted to the promotion of activities in favour of the aerospace industry. The objective of ESINET is the creation of links between national and regional space incubators in the EU Countries as well as in the EU Candidate Members States in order to:

- Promote the creation of new enterprises and incubators: ESINET can allow to exchange on best practices and experiences with other incubators;
- Facilitate the technology transfer between Space and Ground sectors: ESINET can be used as a good motor to extract space technologies from ESA or space industries;
- Help start-up enterprises to have access to finance, to advise them on IPR, networking, marketing and legal matters, and to provide consulting services: ESINET can provide very powerful resources in terms of consultancy and expertise, in order to better help entrepreneurs to build competitive projects (technological experts; IPR experts; marketing studies, transfer feasibility studies ....);
- Establish strategic transnational partnership and networks: ESINET can provide start-up good transnational partnership opportunities for participating for instance to EU proposals;
- Improve the dialog of entrepreneurs with the European Commission and the European Space Agency as well as with national and regional bodies: ESINET can be the optimum interface between space start-up Companies in reason of its critical mass. Moreover ESINET can allow to give space start-up a better visibility as regards to European Space organisations (industrial and institutional);
- Form a critical mass in order to gain visibility as regards to private investors at a European level: through ESINET, incubators, and so their space start-up companies will be able to be visible at a European level as regards to private investors or associations of private investors. Therefore the Space will be better seen a potential market to invest in.

It was the objective of the workshop to provide an overview of the ASLINK experience in supporting aerospace SMEs, highlighting synergies with parallel on-going initiatives as well as introducing the available initiatives in the VI Framework programme of the European Commission. The final goal was to demonstrate how economic and technological intelligence can create the basis for the development of leading edge technologies by a network of aerospace SMEs, attracting contracts and cooperation from large contractors looking for technological solutions for big projects as well as opening new market perspectives for spin-offs in different industrial sectors. All these processes may be pushed and accelerated in a dedicated virtual infrastructure which support companies technically and financially, as well as create a focal point to establish networking and attract contracts from big players.

Agenda

The agenda of the workshop:

- Introduction of initiatives to support SMEs (tools and infrastructures available), ESA
- The Aslink and Scratch projects, D'Appolonia
- The Aero-SME initiative, AECMA
- A virtual infrastructure to support high-tech aerospace SME, T4TECH
- The initiatives for SMEs in the VI Framework Programme, EC
- Round table discussion

Chairman: Mr. Pierre Brisson, ESA, Head Technology Transfer Programme

Technology as a Basis for Collaboration

Rainer Bugow

Prostep AG

The manufacturing industry is forced to cooperate in an international context with different partners in order to achieve shorter development times while expanding their range of products.

Thus the need for engineering communication is increasing and calls for further improvement of technology and infrastructure. This goes along with the need for change and improvement of business procedures as described by the PDTnet project [1]:

New manufacturing techniques and processes in product development require both increasing support for engineering tasks by means of CA tools and PLM.

- The level to which CAD is used in product development in the automotive industry can be assumed to be approximately 90%.
- Technical drawings are being replaced to an increasing extent by digital 3D CAD models as the definitive product documentation.
- The extent to which 3D CAD models are re-used in subsequent CAM, CAE and other planning and logistical processes is increasing.
- Shorter development times force an increase in the parallelization of processes that are, so far, used to be executed sequentially.
- Shorter development times force an increase in the parallelization of processes that are, so far, used to be executed sequentially.
- Improvement in the automation of processes through the use of computer-aided workflow systems.
- Broad use of databases with administration functions to manage product data during the entire life cycle (PDM systems).

New communication technologies, especially with regard to the use of Internet technologies, cause traditional work methods and business processes in the product development process to undergo radical changes:

- The Internet makes the creation of a worldwide data and computer network feasible and makes information available at any workstation, thus laying the foundation for simultaneous engineering.

- The downsizing of computer hardware is now being followed by a downsizing of application software. The Internet browser is replacing complex application software installations at workstations.
- Internet-based solutions therefore also offer an enormous potential for small and medium-sized suppliers of (standard) parts and semi-finished materials. This results in a reorganization of information and distribution marketing channels.
- Networks can be used to digitally simulate the appearance, function / behaviour and manufacture of complex products - at distributed locations, if necessary.

Therefore, attention must be paid in particular to the following general problems:

- Non-coordinated processes interfere with the ability of different companies to work together.
- The implementation of customer-specific requirements ("bilateral optimisation") increases the level of segmentation and results in a loss of synergies, especially with regard to suppliers.
- Legacy data and systems make it difficult to introduce new technologies.
- Competing standards slow down the decision-making process and increase costs.
- Organisational uncertainties have a negative impact on the consistent application of technologies, e.g. relating to the validity of data, security aspects in the network, criteria for evaluating the quality of data.
- The shifting of business processes to a network infrastructure increases the demands made on the integrity and quality of the data.
- The availability of know-how required to master digital business processes is not always distributed to the extent required: well-informed experts from large corporations will be dealing with contacts from SMEs whose core competence is not focused on the area of information technology.

Technology is available to handle these problems, acceptance and success depends on the common understanding of improved business procedures. Standardization of these procedures is undertaken by various organizations. The next step will be the automation of processes driven by workflow systems. The question is how the engineering sector can leverage experiences derived from the Purchasing, Sales, Production/Logistics sector.

References: [1] www.pdtnet.org

MOSAIC E-Net Integration Platform

Reduction of Processing Costs through Electronic Data Exchange

– improving the workflow between the companies in purchasing, sales, production/logistics, engineering and services –

Thorsten Reinheimer

Mosaic AG

In the following contribution, the author demonstrates different application scenarios for the MOSAIC E-Net platform as an Inter-Enterprise integration solution in industry. Considerable effort has been made by industry – in the engineering sector of the automotive industry, for example – to organise and optimise business processes. An essential element is the harmonisation of the exchange of data packages between manufacturers and their first tier suppliers and from these on to second tier suppliers etc. through the web.

The optimisation of engineering processing is targeted at the

- reduction of process costs
- and the
- reduction of the error rate and media breaks.

The solution is an efficient inter-enterprise integration which enables ideally a 100% B2B integration. The tools currently available in commerce and industry differ quite considerably. Whilst a higher level of standardisation has already been attained in commerce through the use of EANCOM / EDIFACT formats, the format environment in the industrial sector is far more heterogeneous. There, SAP-Idoc, xCBL, EDIFACT and other formats exist. There is a demand to exchange these formats via the Internet to keep the costs of data exchange low.

However, the solution described above must fulfil further requirements. Such an integration solution can only gain ground if it is accepted. Acceptance must exist, especially with small and medium-sized businesses. These companies still have great difficulties with the processing of complete business processes or are incapable of it.

In principle, many of the systems cannot support the complete, defined business process. This means; systems which are to be integrated at companies, are only capable of supporting subprocesses.

The Mosaic E-Net integration platform

The Mosaic E-Net integration platform with a hub-spoke approach offers a solution here. The hub is comparable to a central messaging hub for business processes. It has online and offline connectors (spokes). These are used to form links with the partners.

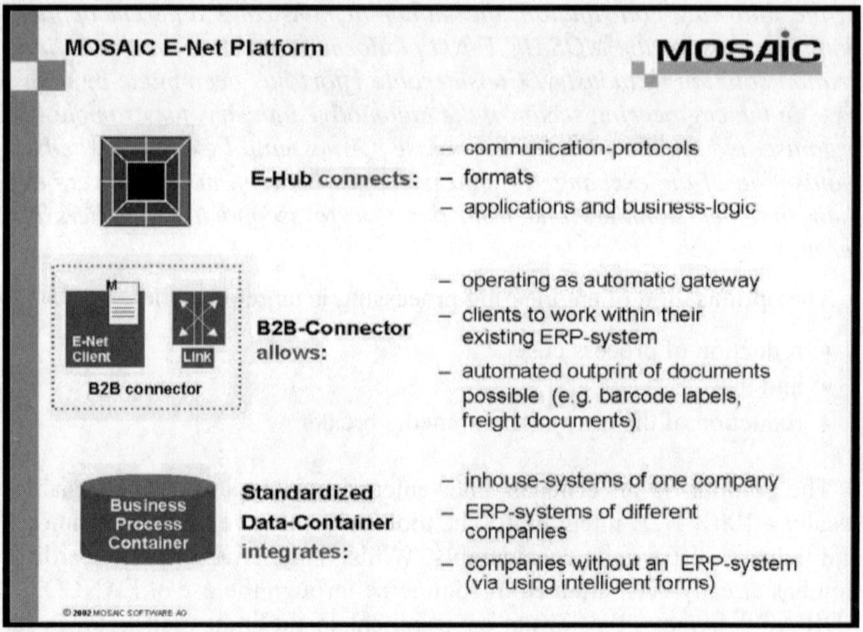

Fig.1: Key components of the platform

An important component of the platform is the business process container. It is a type of virtual file and contains the business process data as an aggregation of messages. This means a sequence of messages which must be processed in the workflow. Furthermore, the virtual file contains a form view for small partners who only support a subprocess through their system or do not support the business process at all, as well as the entire business process logistics.

Moreover, an allocated check of business process rules is performed via the Mosaic platform. This process is explained in more detail later.

The diagram shows the key components of the MOSAIC E-Net platform and its performance in respect of system integration and support of business processes.

MOSAIC E-Hub: this key component combines communication protocols, formats, applications and the business process logic.

B2B connectors:

The platform sports a broad range of connectors. They enable operation from an automatic gateway up to complete manual interaction.

In fully automatic mode, the partners operate within their existing systems. Via the connectors, the automatic output of documents and additional documents such as barcode labels or freight papers etc. can be affected. The connectors also enable a heterogeneous mode of working, meaning that the existing systems support partial steps within a business process, which are then processed in the corresponding in-house system. Unsupported subprocess steps are mapped in the connector in a form.

Business data container: The third key component is the standardised data container, that is the above-mentioned virtual file. Its performance consists of the connection of the different in-house systems at the level of business processes:

- connection between the merchandise management systems within a company
- connection between the merchandise management systems of different companies
  and
- integration of companies without merchandise management systems through intelligent forms which can be processed with the connector.

In Fig.2 the MOSAIC E-Net platform and its functionality are represented as a circle at the centre of which is the MOSAIC E-Hub.

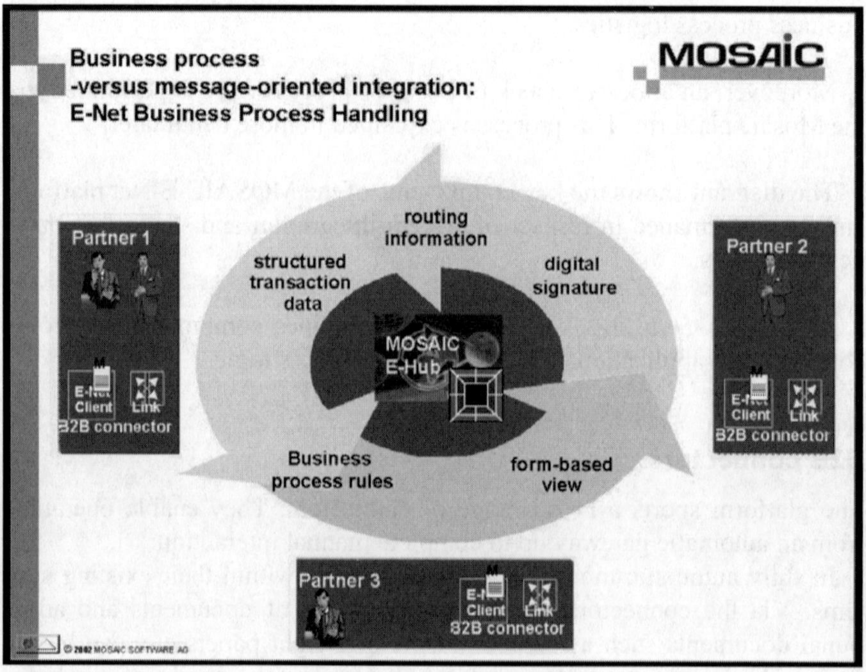

Fig.2: Business process versus message-oriented integration: MOSAIC E-Net Business Process Handling

The business process which is exchanged between the partners in the Internet comprises the structured data, business process rules, the form view, routing information and optionally, of course, the digital signature. The entire business process is encrypted.

This is also what makes up the difference to conventional message-oriented integration solutions. The MOSAIC solution integrates from a business process view.

The diagram demonstrates this using an order as example.

Partner 1 generates a message from his in-house system or manually via a client. The order is taken up and first checked whether it is permissible in respect of the business process.

This means that the system takes an empty template file and tries to file this order in this 'file', according to all the rules deposited in it. This is the actual checking process.

If all rules are fulfilled the order can be accepted. It is then transported on to partner 2 via Mosaic E-Hub who obtains it in the form he requires and can process.

This is an aspect of particular importance to industry: partner 1 must provide all the information such as SAP clients – i.e. internal information – required for ordering which is both valid and can be processed automatically. Partner 2 – a small supplier, for example – cannot process these at all and does not require them. The additional data for the order are also set by partner 1 and transported to partner 2 but only used by partner 3. Partner 2 receives the order in the form he can best process it. He can process it in the client or in his in-house system.

He can supplement the data and then send his invoice, for example. If he imports the invoice into the virtual file the corresponding checking rules are activated. If the checks are successful, the virtual file can be forwarded to partner 3, who could be the central invoice-processing unit. Partner 3 is now capable of fully processing the invoice which partner 2 established with his simple system, as he receives the additional information from partner 1 required within the group (as the SAP clients mentioned previously, etc.), via the business process container, although partner 2 never saw these.

Everyone is presented with the messages or the data in the format they can process and as required for their process step. This is an extremely important mark of differentiation of this data container – this virtual business process file – as, from the technological approach, the business process is viewed and not simply individual messages.

Fig.3: Business process versus message-oriented integration: product life cycle

The diagram shows in which section of the product life cycle we are currently to be found. In commerce, the Supply Chain Management is characterised by numerous structured data within a business process. The process is well defined, i.e. there are EANs and ILNs. Processes can run well on these number systems. The industrial sector displays similarities. The processes are less standardised though. These must be harmonised, however, we are still in an area in which structured data predominate in the process.

If we now go a step towards the engineering sector we are dealing with many unstructured data, from a process view. The industry requires process support in collaborative product development to gain control of the exchange of engineering data.

Exchange of PDM data

What form does the exchange of product-data-management data take? Via a so-called PDM system, the PDM data are exchanged between manufacturers and suppliers. This means that if a manufacturer sends a request to a supplier, an employer of the supplier must select the corresponding parts in the PDM system and initiate the export. In the PDM system, a specific file is generated, for example in Step format, which is then transported further via the network by selecting an address and a data exchange tool – referred to as an ENGDAT package and comparable to the MOSAIC E-Net container, but for CAD and PDM data. It is then received by a corresponding tool, supplied as a PDM file and, generally manually, integrated into the manufacturer's process. Sometimes integration is also effected automatically. In that case, the data are selected via the web (for example, PDT.Net), via a PDM client and a PDM web server, directly by the person requesting it and not by an employee of the supplier. The transfer is still effected as a batch operation, i.e. not online or synchronously.

Process support through MOSAIC E-Net platform

How can the MOSAIC business data container support the process described above?

It could be the case that a TDM or PDM system sends a request to a process supporting merchandise management system. The request is prepared there and transferred to the supplier system via a MOSAIC E-Net container, including the corresponding references, documents, etc. Further processing is effected in the supplier system. If access is permitted, the data are then processed accordingly. The data are transferred to the system so that an ENGDAT package can automatically be exported. This process is referenced from a process view and placed back into the E-Net container. Then, in parallel, an E-Net process container "rolls" back from the supplier to the manufacturer, which references the engineering data and provides process-technological support, so that the data can be imported automatically into the PDM system of the manufacturer. The PDM system then advises the merchandise management system of the successful completion of the process. This could automatically generate a receiving advice, be acknowledged and, for example, released for accounting, via the process-supporting container.

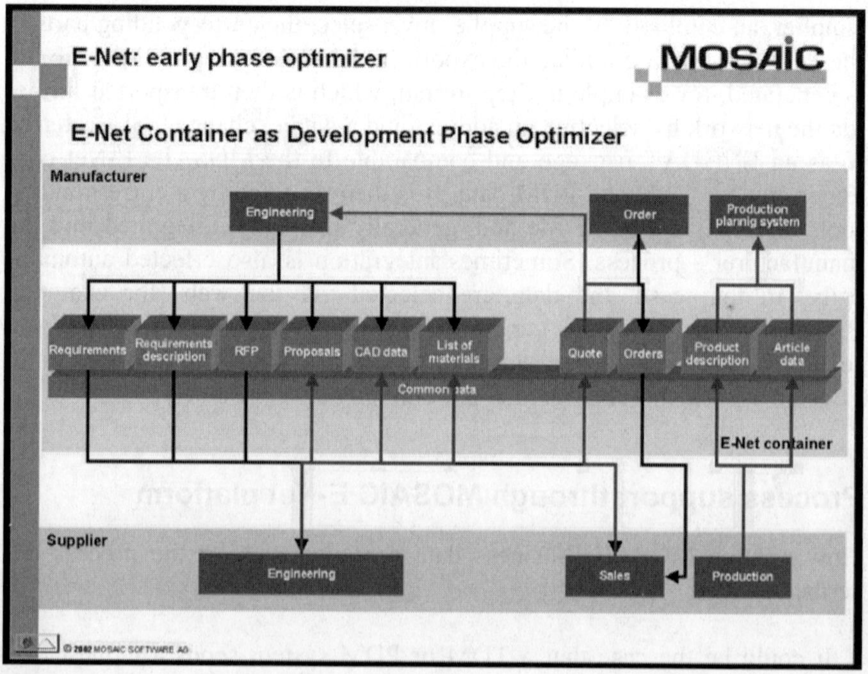

Fig.4: MOSAIC E-Net – early phase optimiser

The MOSAIC E-Net platform could also be used as an early phase optimiser to provide support during the developmental stage. It would be feasible that the engineering department of the manufacturer sends the corresponding requirements or requirement descriptions and similar to the supplier. These data are transferred to the engineering division and to the sales division which process the data correspondingly. The engineering department of the manufacturer supplements data – PDM data, CAD data, etc. – and returns these. Sales could support this process with an offer which is addressed both to the engineering department and to the order management of the manufacturer. This could result in an order from the automobile manufacturer. The process could extend through to the production cycle where product and item data are transferred on to a production planning system. This point virtually represents the transition from the engineering to the production stage.

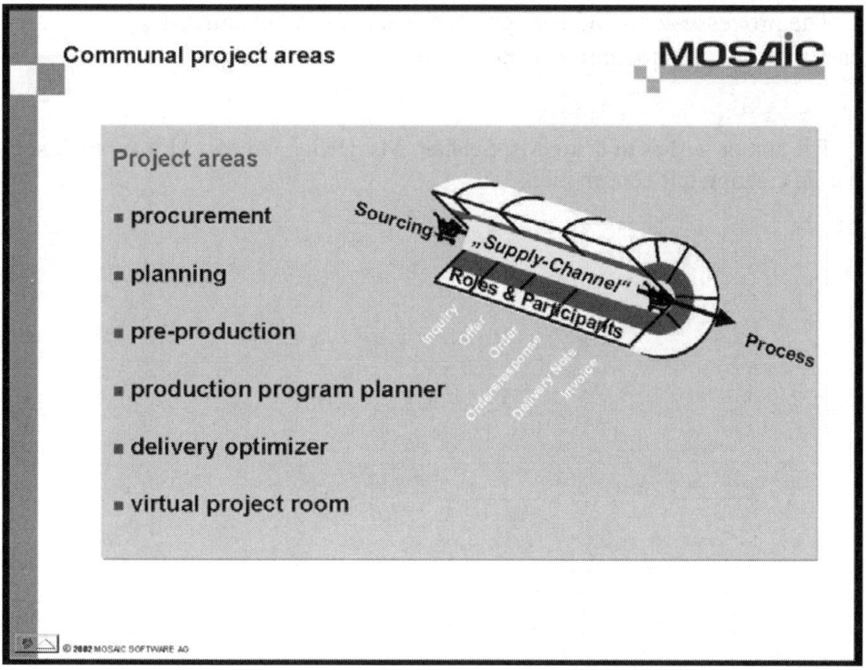

Fig.5: Communal project areas

The MOSAIC E-Net platform or MOSAIC E-Net technology has been used in commerce and plant construction for many years. The business process container guarantees the consistency of the individual steps of the transaction. A successful joint web-based portal for commerce and industry has been available since May 2002, based on the MOSAIC E-Net platform. Connected partners have the benefit of being linked to all participating partners in the B2B through the one connection. The E-Net platform integrates all partners and their merchandise management systems. It ensures that all connected partners operate according to the defined business process regulations. If required an online and an offline solution are available.

The benefits to the supplier are obvious. He is linked to all customers via a single portal. Customers thus achieve a reduction of their investments: their suppliers become 100% EDI capable for them and all business processes accelerate as the potential for errors is practically reduced to zero. With regard to suppliers, this also leads to an increase of customer satisfaction.

The processes existing for commerce and for plant construction can very easily be adapted to engineering processes.

The author wishes to thank Ms. Schäfer, Ms. Heller and Mr. Alex for their support in creating this contribution.

Data Re-presentation as a First Step Towards a Unified IT-Structure

Martin Schottenloher

ewf europe and Mathematisches Institut, Ludwig-Maximilians-Universität München.

The Gap

In many enterprises of the manufacturing industry a gap is felt between development and sales. This phenomenon is not restricted to one aspect but it is related to culture, processes and data formats or – on a higher level – related to people, organization and technology.

On the basis of the installed IT-technology in the enterprise this gap separates the two worlds of EDM/PDM systems and ERP systems (including procurement and service).

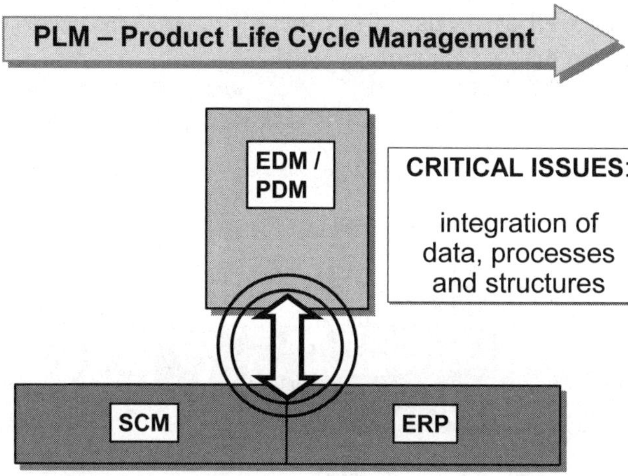

Fig.1: The gap

In both worlds the enterprises have spent substantial amounts of money. In general, the results of the investments are fairly poor, and one essential reason for the resulting disappointment is that the two worlds do not work together.

Although technology is certainly not the main point in general, this gap causes many problems, e.g.

- information is encapsulated in two worlds and communication is affected
- access to data is not given freely
- applications which have to use data from both worlds are difficult to build if not impossible, e.g. for both procurement and service
- field information reaches product development too late

A vision with which to solve this problem of separation is an integrated technology which would incorporate the EDM/PDM world and the ERP world on an equal footing. Unfortunately, there will be no such technology in the near future, although some marketing statements seem to claim that a solution is in sight. Neither the providers of EDM/PDM systems nor of ERP systems are about to develop a system which integrates both.

A Bridge Between EDM/PDM and ERP

In order to enable the collaboration inside the enterprise (and also among enterprises) and to implement important applications depending on development data as well as on data from the ERP base one needs to build a "bridge" between the two technology worlds.

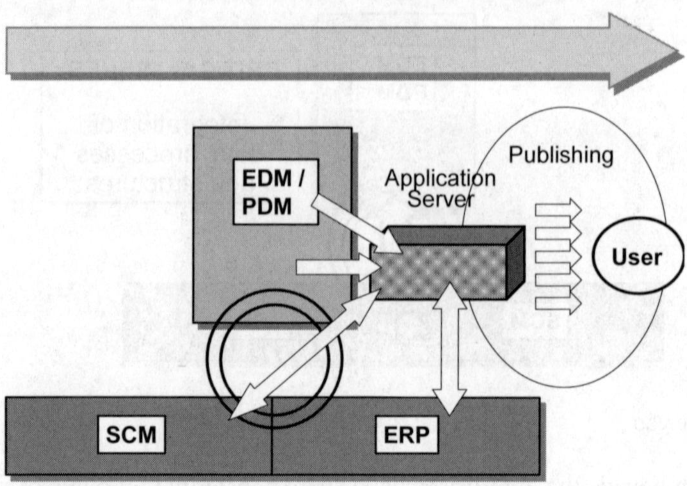

Fig.2: Re-presentation and publishing

Such a bridge should at first concentrate on particular goals, such as bringing together the relevant information in an efficient way for applications in procurement, service, maintenance, assembly, configuration, documentation, sales and more. This means re-presentation of the relevant data. And it should be based on internet and open systems technology.

From the point of view of technology such a bridge can be realised by a software platform in the form of an application server with the appropriate functionalities (Fig.2). The main property is, of course, the possibility of creating the necessary interfaces:

1. Filters for the various CAD formats (2D and 3D) guarantee the fast and efficient transformation of geometry information into a form which can be presented in any browser without the need of a plug-in.
2. Interfaces for ERP systems are comparatively easy to establish. As a rule they have to be customized to the specific installation.

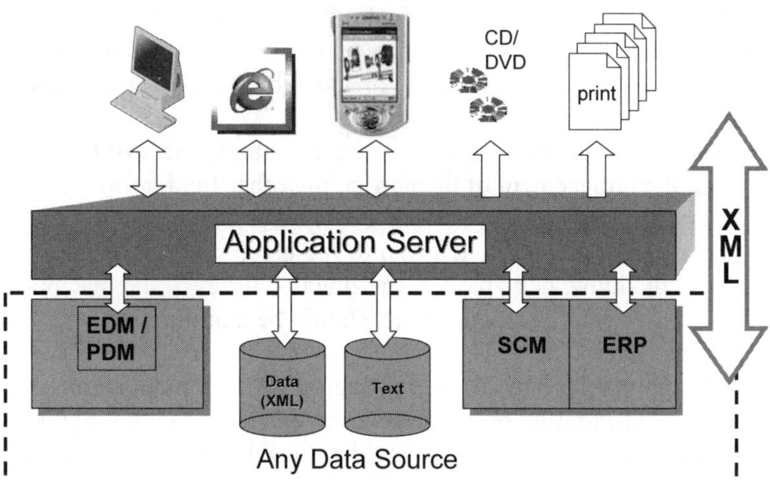

Fig.3: Architecture of the platform

The main additional properties the application server should possess in order to allow important and efficient applications are:

- platform independence
- scalability
- stability
- transparency
- no dependence on plug-ins

- easy creation of applications
- interactive user interface
- secure transactions (ACID)
- availability by web browser through the internet, through LAN, on a PC or a handheld
- availability also as a stand alone application (e.g. CD or DVD)

Such an application server can be built on the basis of Java and writing the relevant applications in XML (Fig.3).

Applications

A relatively straightforward application on the basis of the described software platform is a web based spare parts identification system which serves to cut down the number of wrong orders, and which allows to order directly: The user identifies the relevant spare part (anywhere at any time) by simply finding it in the geometrical presentation (in the browser) of the machine or plant in question. He then obtains all the relevant information about it, e.g. technical information or availability, and he can order the spare part immediately.

An extended application provides the possibility of writing protocols concerning the replacement of the part or any other field information.

In yet another application, the entire technical documentation / communication may be available in addition to the basic spare parts information. In this case, the knowledge base of the functionalities of the machine, construction plans, repair instructions etc. should be administrated in a content management system.

Such an application may be used for corporate communication as well.

Also, the configuration of variants can be carried through virtually in the browser and such an application could be used for product presentation and as a sales tool.

Finally, the software platform can serve as the basis of a collaboration tool within in a company or among several companies.

Competition and Collaboration in Sport

Gunter Gebauer

Freie Universität Berlin

The two principles currently attracting attention in European industry, competition and collaboration, are also the decisive forces in modern sport. In sport, we see an example of how these forces can interact to continuously improve and enhance performance. What Harry Teifel highlights in his contribution – namely, that "collaboration and competition are found in all value chains today and often have significant potential for improvement" – is precisely what unfolds before the eyes of spectators in sports stadiums. The same phenomenon that experts on economic strategy focus on takes place at sporting events, producing striking results: "value adding processes" occur with the objective of aligning "collaborative and competitive forces". Like business enterprises, athletes work towards a "permanent enhancement in efficiency" (Peter Stehle), and "the speed of improvement" (Hans Erich Mundt) gives them a decisive competitive advantage.

In this contribution, I propose that we look at recent developments in the field of high-performance sport that have emerged alongside and in analogy to economic developments. Today the drama of sport unfolds as if on stage – in full public view and with a high level of emotional participation on the part of the spectators – and takes the form of attractive plots and compelling action. Its results are often characterised by their excitement and beauty, qualities which make the comparison with developments in the economic sphere both fruitful and interesting.

In sport there is general agreement on the regulation of competition and general acknowledgement of the final result. Without this recognition of rules, athletic competition would be nothing other than brute violence. What hinders the occurrence of actual physical violence is a player mentality and a longstanding and extraordinarily effective social mechanism that are set up to overcome disorder. At the end point of the competition is the moment distinguishing winners from losers. This outcome is regarded as beyond question and is respected by all. By looking at sport we can differentiate among several kinds of interaction between collaboration and competition, and examine those forms more closely that could be relevant for economics and industry.

At first glance, high-performance sport seems to be nothing other than the purest form of direct competition, and thus the competitive principle seems to play the dominant role. Although the sole objective here is indeed victory, we are not unsettled by competition in this extremely pronounced form. Without being fully conscious of it, we are intuitively aware that in athletic competition, the contenders are neither enemies nor merely opponents, but also partners engaged in *mutually* improving one another's athletic performance.

In answer to the question of what has changed in sport over the last thirty years, we can point to the intensified competitive pressure, which – together with the massively increased financial investment in sport – produces a dramatic enhancement of performance. Furthermore, in football, for example – in contrast to what we might like to nostalgically attribute to players of the past – today's athletes are faster, endowed with better stamina, more advanced tactical skills, the capacity to deal with more complex tasks, and superior abilities to strike the ball and make the shots. Similarly, tennis players today – although they may not be as attractive as the old heroes of centre court – are more athletic; and long-distance runners, too, have improved their endurance and tactical abilities significantly. Today's athletes have achieved a new dimension of playing strength, and the sport heroes of the past – even if they could be endowed again with youth – would not be able to hold their own against them. Certainly the higher degree of competitiveness is one reason for this development, but this is only one side of the coin. After all, competitive sports are still *games*, albeit a particularly intense form thereof, and games are based essentially on collaboration. To attain an adequate understanding of competitive sports, it is necessary to examine the interweaving of competition and collaboration within them.

We can identify two basic forms of interaction between these forces in sport: the interaction within a team, and the interaction between opposing teams. This does not exhaust the topic, however: two further forms can be identified which also prove interesting for the comparison.

1. Collaboration and competition within a team

Achieving a balance between collaboration and competition within a team is not as easy as it sounds. In the top European football teams, much of what takes place on the playing field today would have once been unimaginable: the variety and intensity of interactions, the diverse means of internal communication among players, the interchangeability of roles and positions. One of the newly developed playing strategies, for example, strives to create an overlap situation: one team draws the opposing players to one side of the field and then takes them by surprise, with several players suddenly taking control of the space that has opened up and attacking the opposing team's goal. This kind of innovative collaboration opens up a wide array of new strategic possibilities, and creates new opportunities to catch one's opponents unawares or unprepared. Modern football games make much greater demands upon collaboration than they used to. Thus the basic repertoire of top European football teams always includes an exchange of positions between attackers and defenders depending upon the situation, such that the defenders attack, playing like left or right wing players used to. Such a defender may shoot more goals than a forward, who in turn plays the back of the field and fulfils important defensive tasks and is prepared to move up instantaneously into the midfield to fill any gaps. This form of interaction within a team can be termed *flexible internal collaboration.*

Another remarkable aspect of these new strategies can be seen in the intense competition surrounding the team line-up: only very few players can be sure after one match of being put on the line-up again for the next. On Europe's best and wealthiest teams, the positions considered to be most important are filled by two players. Real Madrid, for example, has two of the best midfield players in the world – Zidane and Figo – competing for the same position; while the stars Morientes and Ronaldo compete for the position of forward. Within every top team, there is fierce competition to make the regular line-up. This we can call *internal competition for positions.*

Thus, the team members are under extreme pressure during the game, but the team itself follows a strategy designed by the trainer without direct reference to the individuals who will later be chosen to play. This means that all players must possess enough flexibility to play well no matter who their teammates happen to be. It also means that each player has to be able to play out of position, to adapt and to collaborate effectively in a variety of positions.

Even when the internal competition among players is at its fiercest, they know that they are part of a team and that if the team does not succeed (if eliminated from the European Champions League, for example) they will be denied any further chance of playing. In this situation, where the players are simultaneously partners and opponents, each one of them must shape the game to his own *and* to his team's advantage. Sociology's traditional models of interaction overlook this requirement when focussing narrowly on the smooth functioning of social intercourse. Many interactions, however, and particularly the most successful ones, are far from the ideal model of the elegant conversation tacitly assumed by these theories. As current developments in football vividly illustrate, the participants in any interaction move within competition-based systems which produce dramatic leaps in performance in a doubly coded field of action: competition in all directions, but collaboration within the team. Here we see two hierarchically ordered levels: the one is competitively oriented and requires that the player distinguish himself from all others without exception, while the second, higher level, requires that the player subordinate himself to a common higher goal and adapt himself to the body of the team while still maintaining substantial personal flexibility. A player must therefore be able to strike a balance between the two demands, thus pursuing his own ambitions but also keeping them in check when the goals of the team take precedence. Thus, no matter how heated the internal competition, there exists *collaboration as the subordination of individual interests to a common goal*.

In sport, competition is a system of performance enhancement, but this is not the only principle at play here, and its function cannot be adequately understood by looking only at the individual player. When we broaden our perspective to include the entire social structure that produces this high-level performance, we see that each individual acts in the context of a larger plan – not an ideal or quasi-divine plan, but one intimately known to the players themselves: it is in the mind of each member of a team.

Thus a team's game results from the planned harmony of the individual wills to compete. This may almost seem like a revival of Mandeville's "fable of the bees", a view of early capitalism that saw a public good emerging from the vices of the competitors. In football, however, the competitors do not act blindly, but possess – along with knowledge of the larger plan – an understanding of the demands imposed upon them by the team. Thus the competitors possess a freedom of will, and acknowledge and accept the team's goals. Even when a player does not make the line-up for a particular game, he grits his teeth and watches from the sidelines, hoping to be granted a better role in the next game. In training he will continue to strive for peak performance and indeed may put in a greater effort than before in order to outplay his opponents, who in turn will strive to achieve ever-higher levels of performance.

2. Collaboration and competition between teams

Until now I have dealt solely with the strategy of an individual team. It is only a small step from what we have observed thus far to what occurs between the members of *opposing* teams. In this kind of interaction as well, an astonishing amount of collaboration occurs. To identify these forms of collaboration, let us look at the entire game as an event that takes place through the efforts of *both* teams. When two opponents confront each other at their peak performance levels, an exceptionally high-level game results. This immediately becomes obvious when one abandons the point of view of the fan and watches with the eyes of the impartial spectator who simply wants to watch an exciting match. Such spectators are enthralled by the performances, the show, the competition, and the interaction of the players on the field: they are the beneficiaries of what in their eyes is an instance of collaboration that offers them something exceptional and justifies their investment of time, attention, and money. High levels of emotional participation on the part of the spectators give rise, in turn, to investments by sponsors, clubs, and television stations. But how does the clash between two opponents generate an attractive product?

The peak performance of just one team is not enough – it takes two teams to make a good game. A strong performance by a team's attackers, for example, is valued highly when they have succeeded in overcoming an opposing defence known to be excellent and tough to penetrate. What is more, when playing against an outstanding team one has to stay particularly "on the ball" – any move forward to attack the opposing goal will likely be answered by a counter-attack through the gap thus created. A fast, sophisticated game compels a team to surpass the speed and ingenuity of its opponents or to thwart their strategy through tactical finesse.

Symmetrically structured competitions are called into being through the outstanding play of both partners. Each team accepts the challenge, the ambition, and provocation of its players by the other team, and incorporates these elements into itself as well. It then behaves exactly like its opponent, but with heightened intensity. It tries to "get into the heads" of the other team to predict how they will move next. This is a form of collaboration that reaches into the furthest corners of consciousness. Georg Simmel impressively described this dual aspect of internalisation and the actions that result from it (Simmel 1903). A strict symmetry of competitive positions leads to an internalisation of the opponent's actions. A competition of this kind is an effort to garner every possible advantage. Along with the physical aspect, it has psychological and intellectual aspects as well: one actually takes the position of one's opponent, becoming so closely tuned in to him mentally and physically that one can anticipate his actions and thoughts and even stay one step ahead.

With the exception of Simmel, social scientists have overlooked this collaborative aspect of classical competition up to the present. A highly successful example of collaboration is the Champions League, in which the best European football teams compete with and against one another. The result is a competition that is fascinating to watch and clearly above the level of the national leagues, and one that involves nearly all of the world's top players. Thus the leading European football clubs (together with the federation) have actually created for themselves new organisational and social frameworks – a clear instance of collaboration that offers distinct advantages to both participants and spectators.

3. Constitutive collaboration

A less visible but in fact more important element is the creation of a framework enabling the game to take place in an orderly fashion according to indisputable, generally accepted rules. These are the constitutive foundations of competition; they define who is allowed to enter the field of competition, how the winners will be determined, what system will be used to classify the players, how the chances will be redistributed at the start of a new season, and how the behaviour of winners and losers in relation to one another will be regulated.

Just how important it is that the losers acknowledge the winners, and Norbert Elias shows that the winners treat the losers fairly, in his historic study of 1986. A regulated, collaborative procedure of this kind does not simply take place naturally but has evolved over centuries in the course of the European civilising process. In 17th and 18th century England, Elias traces the emergence of the democratic process, which we today take largely for granted: the regulated transfer of government power after a parliamentary vote and the regulations restraining winners in their treatment of losers. When it was no longer necessary for English politicians to flee to the continent for fear of their lives after a defeat at the polls, when losers were able to remain in the country instead and continue practising politics, the basis had been established for the regulated process of civilised competition: all participants had understood and internalised the collaborative aspect of competition. This solidly established collaborative mentality was, according to Elias, the necessary precondition for modern sport's emergence in England: once the stage had been reached in which the collaborative aspect of competition was broadly recognised, collaboration could be expanded into all areas of competition as well.

We find examples in which *constitutive collaboration* establishes the conditions for competition throughout recent sport history wherever it was necessary to create a favourable environment for fostering processes of professionalisation, improving the economic basis, and adapting traditional sport structures to modern conditions. This has succeeded wherever we find a market for a desirable product with a commensurate market value: in football, Formula One racing, tennis, the Olympic games, ski jumping. These are only a few cases, however, in comparison to the many other instances of collaboration that remain unsuccessful (for example, in swimming, volleyball, handball, table tennis, hockey, basketball, etc.), where the market shows little interest in the sport no matter how great its success in uniting collaboration and competition.

4. Collaboration and competition in a leading group

We now turn from the world of football to an example of another type of competitive relationship which shows a connection between competition and collaboration of the type one expects today in industry, in which competitors become partners and partners become competitors again (see the presentation by Köpf). These are processes that take place within a *leading group*. In a 10K race (or in the Tour de France), one can see how the field of participants is torn apart when a small group of athletes puts on a burst of speed and breaks away. Within only a few seconds, a whole new structure emerges, with a few bold athletes in the lead followed by all those strong enough to try to close the gap and integrate themselves into this group, that is, to follow its tempo and rhythm. All others fall behind, and the space grows continually larger between those who now constitute "the field" and those in the leading group. While the field is not capable of increasing its tempo, the leading group maintains a pace that leaves its pursuers in the dust. The front-runner fulfils the same task as a pace setter, establishing the tempo and allowing the other runners to follow in his slipstream. Thus the members of the leading group form something like a large organism comprised of a number of individuals. They are united by shared goals of putting the necessary distance between themselves and their pursuers, of covering the extremely long distance at a high speed, and possibly of breaking a record. To do this they need one another – in fact they need solidarity, reliability and the willingness to trade off the exhausting job of leading. This is an example of *collaboration to maintain leadership*.

Collaboration here is only, however, of limited duration. Everyone within this leading team knows that in the last few laps, each of the competitors will work to gain the optimal position to achieve personal success. They must each calculate their strengths in relation to their opponents: the strongest finishers want to make the decisive move just a few metres before the finish line, while the others have to increase their tempo slowly and thus withdraw from the collaborative relationship at a much earlier stage, turning it into a situation of competition. Each instance of collaboration to maintain leadership in sport leads to a *final competition*. While only one of the leaders can win, everyone in the leading group has turned in a top performance, and the losers in this group can find consolation in their outstanding results. Conversely, the winner knows he is indebted to the losers. But while he will never express this nor even be grateful, he will respect them nonetheless.

Here lies a crucial difference between sport and economic activity. Although the two share a strong emphasis on success, this is not the sole criterion for desired action in sport. Even in professional sport, in addition to the aspect of winning there is a second aspect that plays a role for many athletes (although not for all). It is the question of *how* a result has been achieved. Here factors such as elegance, courage, enthusiasm, and personality have a high status. In fact, these are the qualities upon which sport spectators worldwide place great value: they love the football games of the Dutch national team despite the team's inability to qualify for the last World Cup in East Asia. To the spectators, a passionate loser is often worth more than a cool winner. At the point where the emotional involvement of the spectators begins, the analogy between economics and sports ends.

Translated by Deborah Anne Bowen

References

Elias N, Dunning E (1986) Quest for Excitement. Sport and Leisure in the Civilizing Process. Blackwell, Oxford (esp. Introduction by Norbert Elias, pp 19-62)

Gebauer G (2002) Sport in der Gesellschaft des Spektakels. Academia, Sankt Augustin

Simmel G (1983) Soziologie der Konkurrenz. In: Simmel G, Schriften zur Soziologie, Dahme H-J, Rammstedt O (eds). Suhrkamp, Frankfurt am Main, pp 221-242

Goffman E (1971) Interaktionsrituale. Über Verhalten in direkter Kommunikation. Suhrkamp, Frankfurt am Main

Opening Up to Change

People as Critical Success Factor for Collaboration

Eric Scherer

i2s consulting GmbH, Zurich

In industry there is an increasing transformation towards intercompany collaboration and networking of organizations. This process results in an increase of interfaces, a shift of the management function towards operational levels and ultimately a virtualization which entail in a loss of formal structures. As a consequence, the human being becomes ever more important as a stabilizing factor. At the same time, the dynamics and the loss of structures are placing heavier demands on people. This situation can be improved through (1) the application of an action-oriented approach towards workplace design and (2) a strong focus on people during organizational change.

"Change is a constant." The industry is now in a constant state of change. Market pressures, technological advances, new information technologies, organization concepts extolled as the latest fashions, and not least mistakes in management are taking industry on an apparently endless ride on the "merry-go-round of reorganization." Scientists and consultants pounce on listeners with ever new definitions of terminology and modelling methods, and corporate decision-makers promote catchwords such as "reengineering," "virtual organization," or "collaboration" as if they were an established part of the vocabulary and without understanding them correctly.

In many cases, the actual changes in industry stray far from the orderly routes portrayed in PowerPoint presentations by consultants and scientists, and transformation usually comes to an end long before projects have been implemented in full. The new departure towards new forms of organization, opening the organization up to change, is closely connected to the chaos of the everyday of the project and work. At the operations level, the result is gaps and conflicts that can no longer be resolved. There is often nothing that the affected employees can do but shake their heads in amazement and concentrate on the day's matters at hand. The industrial reality may be multifarious, but it is still possible to make some observations that will support or call into question the sustainability of the changes.

A look at the industrial reality

For sure, it is clear that the "dynamics of market conditions" are causing headaches all around. But it is not only market conditions that have changed. In an age when the Internet and mobile communications are found in practically every household, the organizations in which we work are also changing. Before posing the question of how we can optimize our own organizations, it is necessary to examine dominant trends in the world of work more closely:

- Networks and network relationships as cooperative arrangements between companies are increasingly becoming the standard in company organization. Traditional company boundaries are disappearing and being replaced by ever shorter-term forms of cooperation.
- Project work as a normal form of work organization is gaining increasing importance. In the past, the traditional line organization (linear chain of command) provided a framework within which it was quite easy to build identity and find meaning. With project work, identity and meaning must now be derived from an action perspective.
- Non-tradition contractual relationships are becoming more common. The "cappuccino worker", who offers personal qualifications, services and works within a project for a certain length of time and then moves on to another project with another employer, is becoming more of a reality. The task of long-term personnel development is being replaced by the task of short-term integration of employees.
- Responsibility for "employability skills" has shifted to individuals; lifelong learning is promoted increasingly as a goal for individuals as well as companies. A standstill in this regard is no longer an option.

These trends can be observed in many company areas. However, they are particularly apparent in the area of product design and development. Here the individual person is the main production resource. It is this resource that has to be brought into the value-adding process, continuously and in a targeted manner.

The demand for innovation

Companies operating worldwide are being confronted with two contradictory tendencies. For one, products and technologies are becoming more complex and the markets more differentiated. On the other hand, there is great competition pressure, which demands high quality and ever-shorter lead times. Product design and development therefore has to deal with the following strategic demands:

- Shorter times from product design and development to introduction of the product on the market
- Improved quality of products and processes with simultaneous cost reduction
- Implementation of a two-fold globalization strategy, that is,
  - Develop products for global markets
  - Globalize development and production processes.

Many measures are designed to fulfill these demands and in many companies have already resulted in extensive changes to the structural and process organization.

From the strategic demands and the goal of continuous improvement of the organization and the business processes, the following goals can be derived for the organizational level:

- **Improve cooperation and communication in the design and development process:**

  This cannot be ordered "from above" any more than trust can arise upon command. Better communication and cooperation can only be enabled and facilitated; in the end, it must come from the employees themselves. There is no other way to master through division of labor the increasing complexity of production processes and internationalization.

- **Improve planning and coordination of the design and development processes:**

  The necessity to conduct parallel development processes demands improved planning and coordination. Even if the company increases its reliance on employee initiative, it must assure through appropriate measures that the various activities aim in the same direction and accord with the strategic demands.

- **Increase creativity and innovation ability with respect to the product and production techniques:**

  New ideas and their consistent implementation demand appropriate freedoms and leeway and the willingness to take risks. But as a reward they also can produce quantum leaps in technological development and the company's competitiveness.

- **Increase flexibility and adaptability with regard to the organization of work and operational processes:**

  Even the best thought-out form of work organization is optimal only in particular surroundings. Since these change continuously, the ability to change must be anchored within the organization itself. This poses one of the greatest challenges to management.

- **Improve the controlling of process, project, and self for management and employees:**

  This does not mean counting the buttons on the jacket; it means designing a whole new garment. Instead of even more key data and more metrics, what are needed are new methods for gaining experience with new organizational solutions and for evaluating them at every level of action in the company. The more that the organization fluctuates, adapts, and opens itself up to the "chaos," the more imperative self-controlling becomes. There is hardly a function with greater strategic importance in the modern company that is as little developed as this.

Reorganization at the edge of chaos

Increasingly, reorganization processes follow a chaotic pattern. Concepts can no longer be implemented according to traditional reorganization *principles* and in a clearly defined, orderly sequence (analysis – concept development – realization). As a consequence, dynamic, changing structures arise, which from within themselves attempt to adapt to the demands of a turbulent environment [1]. Due to the increasing number of projects, individual projects determine the actual change process to an ever-smaller extent.

Many reorganization plans have no effects at all at the level of operations, or they develop their own intrinsic dynamics that cannot be controlled. Management, as the official vehicle of organizational innovation, is helpless in the face of this situation. As a result, preference is given to simple concepts and centralized control. What often remains of highly ambitious ideas is merely reorganization by "crowbar": massive changes are made to the organization chart – the positions are shoved around, consolidated, or simply eliminated. The context for organization is changed from the outside only; sustainable reorganization from within does not take place.

Virtualization and loss of structure

Although doubts regarding goal orientation and professionalism are legitimate, the drive to reform is widespread in industry. The continuity of the changes is leading to an increasing "virtualization" of organizational structures – whereby this virtualization is a reality that is completely independent of any definition of the concept of the "virtual organization."

The "virtualization" meant here can be described as a "loss of structures." In many companies, structures have become so blurred that even employees at the management level are at a loss to explain their tasks without the aid of the process diagrams and organization charts given to them "from above". A clearly defined management function disappears within multidimensional matrix organizations, multiple reporting pathways, and ever changing intercompany alliances.

Going back to DAVIDOW & MALONE's [2] original definition of the "virtual corporation," we read „To the outside observer, it will appear almost edgeless, with permeable and continuously changing interfaces between the company, supplier, and customers. >From inside the firm the view will be no less amorphous, with traditional offices, departments, and operating divisions constantly reforming according to need. „ (p.5)

Looking at the reality of practice with this theoretical definition in mind, we can see that it is especially this "inner formlessness" that is becoming the new, characteristic feature in industry– despite the fact that it did not arise on the basis of an orderly strategy. If we observe what it is that remains in a company where "not a stone is left standing," what we find are the *employees* – or at least a good part of them. What this makes clear is that the employees in operations are a stable factor in the context of an increasingly "virtual" company. However, employees mostly continue to be "passive" elements at the mercy of the "inner formlessness" and the reorganization merry-go-round. These things not infrequently lead to a loss of orientation and motivation in employees. This gap can no longer be bridged through the efforts of middle management, which traditionally supplied "orientation," helping employees to get their bearings at the operations level.

Structural models predominate

At the latest with the 1991 MIT study "The Machine that changed the world" [3], organizational concepts like fractalization, self-directed team work, the learning organization, and virtual organization became key words in industrial management. They replaced a trend towards automization and changed the focus back to organizational potentials in the company. As a result, process orientation was propagated as proved means of organizational design.

Process orientation is a means to design an organisations structure. It is a framework for organizing workflow and information. Still, industrial firms no longer fit a closed structural model. They are social organizations that, as open systems, behave according to their own dynamics. The dynamics change in a process of evolution that has an underlying psychological and social logic. However, it is no longer possible to explain, predict, and manage the behaviour and development of companies on an exclusively rational basis.

This reveals one of the dilemmas raised by the recently propagated "paradigm shift" from function-oriented structures to process-oriented structures: Although on the one hand the *loss of structures* in favor of dynamics and agility is set as a goal, the means that are propagated to achieve this end continue to be the models that base upon traditional structural thinking [4]. As early as 1959, in his book *The Folklore of Management*, RANDALL [5] described the longing for closed structural models as the "myth of the organization chart," the universal solution. The belief in structural design as a means of shaping the organization goes back to the works of TAYLOR and appears to be an unbroken trend today. A look at the common fixation on process modelling techniques as dominating means of organizational innovation makes this readily apparent.

From lack of concepts to paradigm shift

The fixation on structures, however, continues to produce a stiff and unbending organizational ideal. While it is amenable to control, it is not able to meet the challenges of an ever more dynamic environment. Thus, the *goal* of the new thinking continues to stand in conflict with the *means* of the old way of thinking. Whereas there is a striving towards a change of paradigm, it is never actually achieved.

We need a replacement for the reorganization approaches targeting the creation of new structures that have dominated up to now. The new approach must focus on the ability of the organization to act independently and in a goal-oriented manner and to structure itself. The basis for the new approach is an action perspective. Efforts towards creating a good form of organization will center on optimal execution of the necessary work actions. In the world of practice, the action perspective quite often automatically suppresses the traditional structural perspective. This can be seen ultimately in the routine of a person's daily work and patterns of thought and action, which play at least as great a role in strategic reorganization efforts as new organizational structures. The goal must be to increase the company's sustainability by means of an action orientation applied as early as the initial phase of a reorganization project.

A new look at obvious potentials

There is no doubt about the crucial role of the employee, equipped with intelligence and flexibility, in the area of product design and development. The person is capable of self-organization and is thus in a position to provide impetus to reorganization independently, from "below," thus closing the gaps in reorganization that comes from "above." This ability must be channelled where it is needed and utilized as a potential. This will make it possible to create flexible forms of organization that can adapt dynamically, from an action perspective, to the demands of the environment – the marketplace and new technologies – but also to new strategic priorities. This adaptation will take place even in the absence of a formal project mandate.

"Our employees are our most important asset," proclaim innumerable companies. However, the list of expectations placed on employees continues to expand, and many a frustrated manager yearns for new employees, having the feeling that the old ones can apparently not change.

In many companies, the management is at a loss to deal with this situation, responding with a regretful shrug of the shoulders and vague references to the much-cited "soft skills." It is apparent that many companies have stopped midstream, so to speak. Employees receive various kinds of training, but anything beyond that falls into the category of "soft factors." Not infrequently, institutionalized human resources and organizational development are sacrificed for cost reasons. Management is expected to fill the resulting gap with their "healthy" management "commonsense."

VANSINA & TAILLIEU [6] provide an apt description of the current attitude towards people: they find that „nowhere in the now abundant literature on reengineering is there an indication that attention is given to the legitimate needs of the people employed. People appear in various categories: leaders, redesigners, people who ought to be flexible, committed to continuous learning or as subject to be committed, to develop shared values and goals; and as an average of 20 percent who need to be moved out."

The capitulation that is evident here and that takes place day by day over and over again, is as striking as it is characteristic. As soon as something is as vaguely defined as the human being is, it is beyond the reach of the traditional tools of organization design. And unjustly so, for even in Taylor's day, researchers and practitioners concerned themselves with the interaction among people, technology, and organization. While the focus in research and in practice was on production work for almost a century, there now is – and needs to be – a noticeable further development to those company areas that deal with product and process innovation themselves.

From "soft factors" to "hard methods"

This much is clear: action-oriented organizational design makes it possible, through targeted methods and criteria, to shape those aspects that up to now have been treated as soft factors. The area where design can be applied turns out to be the *work task*.

Structure and task

Close examination reveals that the structure of every organization is more or less dynamic. Structure itself arises through the activity within an organization and thus bases upon the actions of the actors. If actions are terminated, the organization ceases to exist. Since actions are not always the same, and since they of themselves change in form and type through time, the structure of a system can be defined precisely only for a given point in time. It is, however, futile to attempt to reach an exact description of the state of a system, for the structure of the system changes rapidly. Change occurs both through implicit self-structuring and through explicit reorganization measures, the processes that always occur in the system and the environment. In this sense, instead of "structure" we can speak of the permanent "structuring" of an organization. As WEICK [7] remarks, structures are both the means and results of actions. These actions are executed by the actors in the system in order to fulfil the tasks assigned to them. This means that the tasks that are assigned to the actors in the system produce actions, which themselves in turn serve fulfilment of individual tasks. The structure of the system is thus not defined exclusively by the formal structural and process organization of a company, but instead arises originally from the *definition of the work tasks*. The operational execution of the work tasks is what actually generates the real, dynamic system (see Fig. 1).

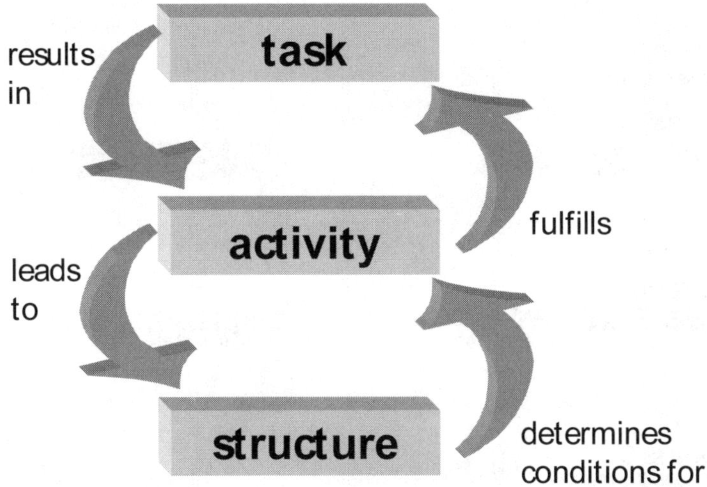

Fig.1: Interaction of task and structure.

In other words, the essential *design aspects* are *structure* and perceived *work tasks*. Shaping the structure is at the same time both a condition of and the result of the design of the work tasks. While there are no standardized solutions, it is possible to derive some rules of systems design. Here the goal must be to design a system that is flexible, adaptive, and robust.

In this connection, FRESE [8] uses the terms "coordination efficiency" and "motivation efficiency." Coordination efficiency serves mainly as a measure of structural design, whereas motivation efficiency can be utilized as a measure of task design. In industrial practice, much more weight is placed on coordination efficiency than motivation efficiency. But to achieve a sustainable form of organization, it is necessary to assure a certain balance between the two.

Motivation efficiency and task design

As employees perceive it, total processes are made up of a number of operations; they are familiar with the execution of those operations (Figure

2). The sum of the operations across all possible business processes forms an employee's realm of tasks, or task area.

In order to incorporate the employee into the process organization effectively, it is necessary to ensure that the employee's task area is positive and challenging. Tasks must be designed so as to allow the employee to develop motivation and the willingness to take on responsibility and to gain knowledge and experience.

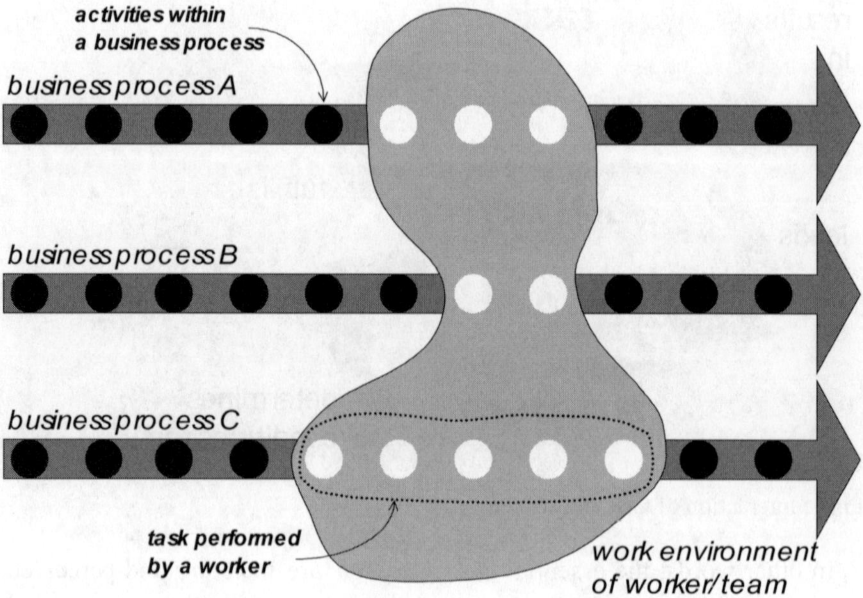

Fig.2: Relation between business process and realm of tasks [9].

Work tasks should be designed according to work criteria and human-oriented criteria (humanization of work). This makes it possible for the person's potential to be utilized in a meaningful way and, at the same time, produces a work environment that motivates people. In this way, people develop intrinsic motivation, motivation from within themselves, which is not the same thing as extrinsic motivation in the form of compensation for performance or special bonuses, for example. Figure 3 shows human-oriented principles of increasing the motivation efficiency of the organization.

Design feature	Expected effect	Realized through
Comprehensive, "whole" tasks	➲ Employees recognize the meaning and value of their activity. ➲ Employees receive feedback on their own work progress from the activity itself.	• Tasks that contain planning, execution, and controlling elements. The employee has the opportunity to check if the results of his own work accord with the demands set.
Variety of demands	➲ Employees can utilize a variety of abilities, knowledge, and skills. ➲ Monotonous tasks are avoided.	• Tasks that make a variety of demands on physical functions and the perceptual senses.
Opportunity for social interaction	➲ Problems are mastered cooperatively. ➲ Mutual support is an aid to coping with stress.	• Tasks that facilitate or require cooperation.
Autonomy	➲ Strengthens self-esteem and willingness to take on responsibility. ➲ Allows employees to experience that they are not without influence or significance.	• Tasks that give the opportunity to plan and make decisions.
Opportunities for learning and development	➲ Keeps up all around mental flexibility. ➲ Professional qualifications maintained and increased.	• Problematic tasks that for their solution require employees to utilize their existing qualifications and to expand their qualifications.
Elasticity with respect to goals and stress-free regulation	➲ Hinders inappropriate work overload. ➲ Creates some space for thinking things through and self-chosen interactions	• Creating buffer time when deadlines are set
Sense and meaning	➲ Employees feel that they are participating in manufacturing products that are useful to society. ➲ Ensures alignment between individual and societal interests.	• Products that are not doubtful in terms of usefulness to society. • Products and technologies that can be checked and assured as non-damaging to the environment.

Fig.3: Features of task design that create a task orientation and foster intrinsic motivation [10].

Competence and leadership

All employees must participate in making the social system dynamic. What is required is a problem awareness that is shared by all parts of the company that are involved. This is just as important for operations employees as for strategic company management. In addition to work competence, the demand for social competence and entrepreneurial behaviour is of particular significance (Fig.4).

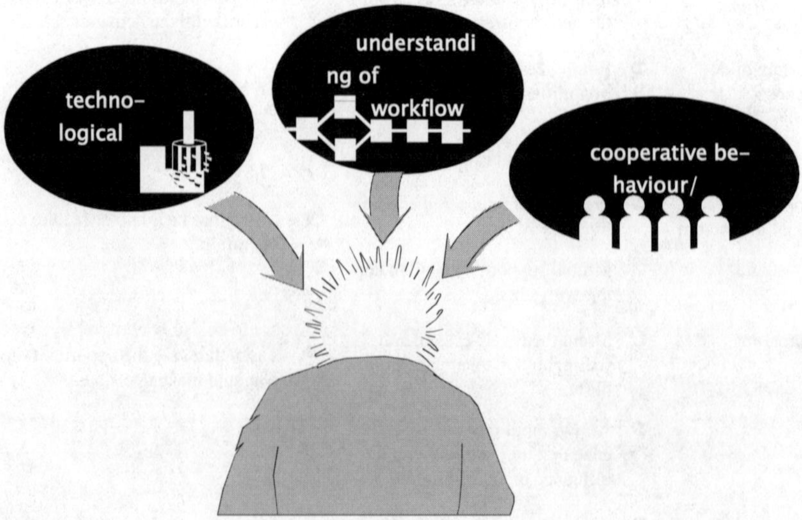

Fig.4: New demands on employees.

To ensure that the striving for qualifications as well as entrepreneurial behaviour will be lasting, it must be complemented by an appropriate compensation system [11].

When the opportunity for self-direction is increased, it becomes necessary to re-examine the role of leadership. Behaviour patterns rooted in power strivings and control have to be abandoned. Strategic company management has the task of assuring operations and the availability of all required resources. To this purpose it will have to create leeway and redundancies in order that a dynamic organization can develop at all. At the same time, company management also has the responsibility to develop the strategic goal system of the company from the top and to align it with the individual, "de facto" goal system at the level of operations. The compensation system has increased importance here as well; it should be constructed in a way that the operations level itself automatically follows the goals that the compensation system demands and promotes. The key to success is the flexibility of company management to negotiate specific solutions jointly with the operations areas.

Goal conflicts and dilemmas

It is usually not difficult to reach agreement over the formulation of goals. However, implementation of the strategic goals results inevitably in goal conflicts and interest conflicts that hinder any "one best way." For this reason, the organization will have to be equipped to walk a number of tightropes simultaneously. It will have to master the following dilemmas:

- **Creativity versus plannability:**
  Creativity needs the freedom in which to develop; planning, by necessity, entails restricting this freedom.
- **Routinization versus flexibility/adaptability:**
  Routinization means standardizing procedures to make them more effective and more reliable; but a flexible organization, however, depends upon the ability to continuously modify the rules.
- **Shorter lead times versus quality improvement:**
  The shorter the lead times for development, the fewer the opportunities for early error detection and quality control.
- **Local optimization versus overall optimization:**
  Local efficiency develops best the more that entrepreneurial autonomy is granted to the decentralized units. The efficiency of the total organization cannot be increased significantly without the contributions of more efficient units. However, at the same time the economic interests of the local units can endanger the efficiency of the organization as a whole.

What is more, the relationship among these goals and priorities changes continuously in dependency upon market conditions. An organization that wants to achieve sustainable effectiveness and efficiency must become aware of these dilemmas and find creative ways to master them in each situation. Static organization models and thinking in terms of "organization structures" are a major impediment.

Levels of sustainable organizational design

What exactly is the meaning of sustainability and "organizational innovation as a core competence"? In the first place, it means self-observation. In other words, there must be continuous analysis and reflection upon one's own actions, the measures taken, and the rules set up – and this at four levels at the least:

- Structural organization (structural and process organization): Structural and process organization form the "structure" of the organization, which can be depicted in diagrams. Traditionally, reorganization projects in industry target the design of the structural organization.
- Operations organization (work organization): How can you divide the endlessly complex process of developing a motor vehicle into operational steps that individuals and groups can execute most efficiently and effectively? And how well are these individual operations in fact performed? The real organization of work at the operations level frequently diverges greatly from all structural dictates. In the Taylorist organization, for example, efficiency was often only achieved because many people took over responsibilities that transcended their narrowly defined tasks. Conversely, reorganization today often takes place "on paper" only. Sustainable design of work organization means bringing strategic priorities "to life" or changing them in such a way that they can become filled with life.
- Changing organization: Organizations change, whether desired or not, in dependency upon external conditions and internal actions. Strategic control of this process (or at least the task of having the greatest possible influence on the process) has up to now been assigned mainly to top management. In an adaptive organization, however, it becomes the task of each and every employee.
- Company culture and individual motivation: Company culture is decisive when it comes to employees' motivation, trust, and willingness towards cooperation. But you cannot "dictate" culture. Since many managers today persist in clinging to this illusion, there is stuff aplenty for lunchroom ridicule and cynicism. Corporate culture is not a "tool," but instead the totality of practiced trusting relationships, type of interactions, and practiced values that characterize a company. If you want to improve the culture that is actually practiced in the company, you must improve the conditions under which the development of trust is fostered.

Consequences and side effects

Anyone wanting to design organization is a sustainable manner must be able to distinguish between consequences and side effects. Beyond the desired effects, every measure will have a number of other impacts. Organizations can be "blind" to such effects and thus ignore them (for instance, as when Total Quality Management (TQM) exponentiates into yet another form of bureaucracy). Managers do not want to be reproached for "errors."

Their competence is, after all, still measured according to how "efficiently" goals are met – or made to look good through citing metrics – and not in terms of the extent of their contribution to organizational innovation. Thus, sustained and conscious organization design means being aware of and monitoring side effects, not ignoring them. The concepts being followed and the implementation of those concepts must be continuously measured against the original goals and intents, and adapted and corrected where necessary. The following are some of the aspects that may be taken into consideration:

- **The relation between formal and informal organization:**
  How do organization chart and process descriptions relate to the real patterns of behavior, or action requirements, in the reality of daily work (for example, the "not my responsibility" syndrome)? Is there a contradiction between management's official goals and what employees perceive to be management's "hidden agenda"? How does management view employees? How is this translated into their actions and directives (demands versus reality)?

- **Knowledge management and barriers to learning:**
  Where do conflicts of interest hamper cooperation across the functions (for example, the "not invented here" syndrome)? Does the culture of the different professions hamper interdisciplinary exchange of knowledge (for instance, conflicts between product designers and management experts)? Where do unspoken rules and standards hinder information sharing within work groups that are composed of different company areas?

- **Team thinking:**
  To what extent is the performance of employees actually viewed and evaluated as a team result and not as individual performance? Do developers have the freedom that they require in order to achieve creative achievements for real innovation? Are developers truly embedded within the cooperative efforts that are required for rapid implementation?

- **Pressure to conform:**
  With group work (such as in project groups), what can effectively counteract too great a tendency towards "harmony" in the group or "group think"?

- **Motivation and disinterest:**
  When changing the organization, how successful is the tightrope walk between strategic goals and consideration of the employees' interests

and abilities? What can be done to counteract disinterest on the part of employees?

Conclusion

It is crucial to always take into account the fact that organization occurs in the mind. Action-oriented design of the organization that attends to the issue of motivation efficiency right from the start of the change project will always lead to more sustainable success than approaches that target structures alone.

References

1. Schultz-Wild, L., Lutz, B.: Industrie vor dem Quantensprung. Eine Zukunft für die Produktion in Deutschland. Springer, Berlin 1997.
2. Davidow, W.H., Malone, M.S.: The Virtual Corporation, Structuring and Revitalizing the Corporation for the 21st Century, Edward Burlingame Books/HarperBusiness, NY, 1992, p. 5.
3. Womack, J.P., Jones, D.T., Roos, D.: The machine that changed the world. Rawson Associates/Collier MacMillan, New York/Toronto 1990.
4. Roth, S.: Neue Denkweisen statt wechselnde Konzepte. In: Maier, P. (Eds.): Reengineering – Fluch oder Segen? Gabler, Wiesbaden 1997. S. 37-68.
5. Randall, C.B.: The Folklore of Management. John Wiley, New York 1997 (Erstauflage 1959).
6. Vansina, L.S., Taillieu, T. (1995): Business process reengineering or sociotechnical system design in new clothes? In: Woodman, R. W., Pasmore, W. A. (Eds): Research in organizational change and development. Volume 9. JAI Press, Greenwich. p. 81-100.
7. Weick, K.E.: Technology as equivoque: Sensemaking in New Technologies. In: Goodman, P.S., Sproull, L.S. (Eds.): Technology and Organizations. Joosey-Bass, San Francisco 1990. p. 1-44.
8. Frese, E.: Grundlagen der Organisation. 6. Auflage. Gabler, Wiesbaden 1995
9. Scherer, E., Zölch, M.: Design of activities in shop floor management: A holistic approach to organisation at operational business levels in BPR projects. In: Browne, J., O'Sullivan, D. (Eds.): Reengineering the Enterprise. Chapman & Hall, London 1995. p. 261-272.
10. Ulich, E.: Arbeitspsychologie. 3. Auflage. vdf/Poeschel, Zürich/Stuttgart 1994. S. 161.
11 Ulich 1994. S. 449-462.

The Boundaries of "Marketisation"

The Myths About Entrepreneurial Employees

Stefan Kühl

University of Munich. Department of Sociology

Businesses' internal control over market mechanisms is propagated using concepts of the "one-man-enterprise", the "intrapreneur" and the "self-Ltd.". At the same time, individual employees are understood to be autonomous entities within the company. This article deals with the myths in management literature about intrapreneurs. It is shown that the contradictory environmental demands within a company are only dealt with by sharing the work out amongst different departments to a limited extent, and are more commonly passed on as paradoxical behavioural demands on the so-called "intrapreneurs".

1 Introduction

In neo-liberal theory, markets are seen as a particularly rational form of reciprocal relationship. While terms such as "hierarchy", "bureaucracy" or "control" have overtones of inefficiency and arbitrariness, markets are considered a particularly fair and effective form of distribution of goods. It is implied that corporate division of labour can be particularly well organised in self regulated markets and that central authorities such as the state should keep out of market processes as much as possible.

For some years the market has been increasingly popularised as a central instrument for the *internal* organisation of companies.[1] With the use of key words such as "organisation networks", "marketisation", "market controlled decentralisation", "internalisation of the market" and "strategic decentralisation", the competition between businesses of varying sizes has been depicted as an instrument for the internal coordination of businesses.

[1] The discussion about internal markets dates back to the early 20th century. The principal first won recognition in general management discussion through concepts such as Lean Management and Business Process Reengineering.

The management thereby simulates a form of capital market in the relationship between the top of the company and decentralised entities and promotes the formation of intragroup work, management, resource and product markets.

While in the beginning the principles of "marketisation" almost exclusively referred to the relationship between company headquarters and organisational subunits such as plants, profit centres or partly autonomous groups, marketisation has since then been increasingly propagated as a model for the relationship between businesses and their employees and for that between workers. Using concepts such as "intrapreneur", "one-man-enterprise" or "self-Ltd.", the members of a business are depicted as no longer seeing themselves as company employees- as organisation men or corporate men- but rather as quasi "enterprisers within the enterprise".[2]

With these concepts it is proclaimed in management literature that entrepreneurship is required from every employee. After all, according to this logic, it is the company's aim to undertake something rather than leaving it undone. For far too long "controllers, royal vassals and nitpickers" have concerned themselves with ensuring that employees only do what is allowed according to their job description. It is about time, according to the management rhetoric, to condemn the "feudal economy" within businesses and to make sure that employees no longer work beneath "aristocratic" bosses, but rather operate as autonomous enterprisers within the enterprise.[3]

Given the prevalent euphoric- sounding management literature about "intrapreneurs", "one-man-enterprises" or "self-Ltds." it would be a natural reaction to dismiss the propagation of the "enterpriser within the enterprise" as the latest colourful management balloon and to declare these "enterprisers" to be merely fashionably dressed employees (and thereby

[2] The fundamental issue is older. Even Schumpeter (1926, p.111) took the concept of the enterpriser from the unity of capital property and function of leadership. In his opinion managers could also act in an entrepreneurial way as "employees" of an incorporated company.

[3] Up until now there has been hardly any scientific thought into how concepts of the "one-man-enterprise", the "entrepreneur" or the "self-Ltd." could affect the functioning of companies. Sociological, ergonomic or business management studies take either profit centres or partly autonomous groups of larger decentralised entities into consideration or else predominantly quantitatively arranged studies about subjective changes in work demands are carried out, which are however solely restricted to the functionality of organisations.

explain the lack of research into the concept of the intrapreneur). In contrast to profit centres, which have in the meantime been employed as an internal structuring principle in many businesses, there have been few examples until now (and also in scientific thought) of companies for which the "intrapreneur", "one-man-enterprise" or "self-Ltd." shapes their appearance. This gradually indicates how a "conglomerate of internal enterprises" can function and to what extent it differs from a "company with employees".

Regardless of the still low diffusion rate of the concept, the vision of the intrapreneur as a model is in the meantime influencing a succession of reorganisation projects, the consequences of which can already be observed in the businesses. The concept of empowerment is orientating itself in much the same way as the concept of target agreement on the model of "intrapreneurship".

Below, I take the concepts of intrapreneurs, one-man-enterprises and self-Ltds. to their word in order to bring out structural problem areas which can develop as a result of selective introduction of entrepreneurs into the company. Three myths of management literature, "adoption of intrapreneurship at all levels", "intrapreneurship makes employees the new leaders" and "integration of employees into the business through intrapreneurship" serve alongside as a starting point for my sociological line of argument.

2 Myth: Intrapreneurship will be introduced at all levels of the organisation at the same time

The perception in management literature is that marketisation will be implemented at all levels of a business- at that of profit centres, groups and teams as well as that of individual employees. This is the case, for example, in one of the leading management books, in which the concept of the "enterpriser within the enterprise" is propagated so that the "entrepreneurial understanding, thinking and behaviour of all employees" is questioned (Warnecke 1992, p.226). The head of IBM, Lou Gerstner, naturally takes as a starting point the fact that his colleagues enjoy winning- "as individuals and as a team" (quoted in Schmidt 2000, p.38).

The scientific analogy which is used for this acceptance of market-moderated principles at all levels of organisation is that of the self-organised fractal. The idea of the fractal- orientated management doctrine is that decentralised entities should adapt themselves to the ever-changing general framework through self-organisation. In this process of self-organisation, according to this presumption, profit centres, groups and teams as well as individual employees become similar in the way they function. Every fractal, and consequently every position in the firm, should function in the same way as the whole business. This results in the most widespread efficiency and independently solved task possible. The coordination between the individual fractals takes place through market-moderated organised service relationships.

2.1 Not everybody can be an enterpriser – contradictions in the concept of "enterprisers within the enterprise"

In management literature, the conception of the "intrapreneur", "one-man-enterprise" or "self-Ltd." is construed as a "win-win situation". This takes as a starting point the fact that all members of an organisation can profit from the implementation of "entrepreneurship in the enterprise", as long as they willingly stick to the principles of entrepreneurship.

This, however, overlooks the fact that the various "intrapreneurs", "one-man-enterprises" and "self-Ltds." within an organisation are, as a rule, so interwoven with one another that they curtail one another's freedoms. The freedom of one, according to an old view of social science, is the insecurity of the other.[4] Even one of the founders of business management, Erich Gutenberg, asserted that the gain of freedom, and consequently entrepreneurship, can only be bought by the relinquishment of freedoms in other areas of the organisation (cp. Gutenberg 1983, p. 273).

The fact that ideas about intrapreneurs do not gain acceptance amongst all employees of a company without problems becomes particularly evident when people involved at different levels of the organisation are concerned. Degrees of freedom on the one level in no way determine equal independence on the other. Self-control on one level can on the contrary mean external control on the other. These ideas are developed below by

---

[4] Luhmann, for example, takes up an old argument of Parsons´. Parsons objected to attempts to democratise all institutions and to leave to the individual all decisions pertaining to them, arguing that this would only work if the individual people involved had no relationship with one another. The moment that assumptions of social interdependence between individuals are acted on, individuals can no longer make decisions which concern them because the freedom of one becomes the insecurity of the other.

means of the interplay between three levels in decentralised enterprises-the profit centre, groups and individual employees.

Firstly, the interplay between profit centres and employees: the appointment to head of the profit centre is tied to an increase in power when compared to a classic intermediary management role. While a head of department in production must always coordinate with colleagues from departments for quality assurance, purchasing or construction, the head of a profit centre has, as a rule, all the important functions incorporated in his domain. Only by taking on the overall responsibility for quality, stocks, appointments, employees, costs and throughput time can he be made responsible for the market success and the resultant win or loss.

This development from "small enterpriser" to management level of the profit centre does not necessarily lead, however, to his colleagues also becoming "intrapreneurs". The opposite is in fact partly the case. Just returning to the concept of intrapreneurs, authoritarian management structures can develop in profit centres. The profit centres can become small princedoms with very harsh managers. It is precisely the pressure on profit centre managers from company headquarters which can lead to them feeling compelled to directly interfere in the work of their colleagues and to limit their autonomy.

Secondly, the interplay between profit centres and teams: in the literature, a distinction is made between a strategic market- orientated decentralisation and an operative decentralisation put into practice in a work process. The interrelation between operative and strategic decentralisation is not so uncomplicated as is commonly suggested. The management of a profit centre is passed on to a manager who will act as an enterpriser within the enterprise.

Included in this entrepreneurship is his ability to make decisions about the internal structure of the profit centre. If the company holding decides that employees should be motivated to greater, more entrepreneurial thinking by the introduction of partly autonomous groups, this can be perceived by the profit centre manager as a restriction of his autonomy. He may make reference to the fact that he can only be an entrepreneur when he simultaneously stipulates how his profit centre is organised. This can lead to the manager deploying and cancelling teams almost at random and justifying this action with reference to his entrepreneurial autonomy.

Thirdly, the interplay between groups and employees: in the group work literature it has long been taken as a starting point that the options open to individual employees are broadened through group and team work. Group work particularly is understood as a consequent continuation of job expansion. But even in the early eighties it was critically noted that partly autonomous groups would not necessarily mean freedom of authority for their members. It was argued that the autonomy of a group of people cannot be equated with that of an individual because the group members can repress one another.

Many empirical studies have since shown that the pressure on individual employees within partly autonomous groups compared with classic tayloristic labour organisation commonly increases rather than decreases (see an overview in Fröhlich/Pekruhl 1996, p. 112 et seq.). A dynamic frequently develops in groups which does not lead to self-development and social recognition for its members. On the contrary, group work is commonly felt to be a "source of power-exertion and aggression, contempt and discrimination". The pressure is therefore perceived by the group members as even stronger because deviation from the group norm can be harshly sanctioned. While managers must as a rule keep to the company's official catalogue of sanctions (caution, withdrawal of salary, dismissal), groups can impose their sanctions on deviant group members much more extensively. The sanctioned group member has very little opportunity to complain to the manager because the power relationships within groups are very diffuse and attracting attention from outside the group to conflicts within it is often negatively sanctioned as an uncooperative deviance from the unwritten rules of the group.[5]

2.2 Internal conflicts and informal processes of attenuation

What happens nowadays when a business uses concepts of the "one-man-enterprise" to try and establish autonomy in individual employees on all levels? I have observed the formation of new conflict areas for which,

[5] The American sociologists Smith and Berg (1987) speak in light of this tendency about the "paradox of regression in groups". Every process of group formation requires each individual member to restrict their identity and individuality to only a few parts. Groups do not gain their strength through their members contributing as whole people but on the contrary through the retraction of particular areas of individuality on the part of each group member. The process becomes a paradox because groups can only allow their members freedoms when they have beforehand limited their individuality in order to successfully produce a shared group effect.

however painstakingly, attenuation mechanisms must be found within the organisation.

In the enterprises of the welfare state in the second half of the twentieth century, relationships appeared to be unambiguous. As soon as I was recruited by a company and began to work there I became "one of the family": I worked for "Chrysler" or for "Microsoft". With all internal career struggles and conflicts in everyday working life, employees could refer to a stable friend-foe image. The businesses which operated in the same market were seen as the competition. Internally, competition was eliminated and a shared company identity was maintained: "us vs. the rest of the world".

In businesses which try to control internal processes of market and competition mechanisms using the concept of the "intrapreneur", this clearly defined friend-foe image becomes blurred. It has already been generally asserted in the most recent studies of decentralised organisations that a social-Darwinist climate has developed in these organisations. The main social-Darwinist feature is illustrated, as a rule, by conflicts at the same level of the organisation: business divisions and profit centres compete for tight resources. Different departments of the same company find themselves in competition with one another. The internal competition is to an extent stronger that the competition with business rivals. On the employee level the "doers", "day managers" and "what does the world cost- types" compete with one another and the climate amongst colleagues becomes tougher.

This competition between entities at the same level is added to by the competition between different levels through the introduction of "intrapreneurs" on all levels of the organisation. The profit centre manager is not only in competition with other profit centre managers but also with his colleagues, who try to establish themselves as "enterprisers within the enterprise". The group enters into competition with employees who try to establish themselves as "intrapreneurs".

In decentralised businesses this does not generally lead, however, to an outright fight of "everybody against everybody else". This relates less to management endeavours towards (re)integration of employees (keyword: business culture, corporate identity), but rather to the informally developing mechanisms for regulation of conflict. It can be observed that the concept of the "intrapreneur" undermines the perception of partly autonomous group work which has become popular in management in the last few years. A decision against collaborative work in permanently fixed groups is made more unconsciously than consciously because of the orientation towards "demanding entrepreneurship from every employee". Permanently changing teams are more likely to catch on than fixed groups with clear-cut tasks and members due to the propagation of the concept of the "intra-preneur".

The realisation of entrepreneurial behaviour at all levels of the company appears to be an illusion. Even when the intrapreneur is propagated as a central structural feature of the company, informal and often unconscious processes develop with which the company employees defuse the threat of competition. The demand for entrepreneurial behaviour at all levels of the organisation can be maintained in management discourse but is replaced in everyday working life by a more realistic form of labour organisation less likely to cause conflict.

3 Myth: employees become the new leaders in businesses through the implementation of entrepreneurship

At first glance it may seem paradoxical to market theorists for managers to call for the internal market- it would eventually lead to the abolition of their profession. Following the logic of a liberal market theory, central leadership is no longer needed by the market. The hierarchies would erode. The many small "enterprisers within the enterprise" would become the "new leaders" in the enterprise and would oust the bureaucratically organised managers.

But evidently the impulsive and frivolous act of public suicide on the part of managers is not taking place. Despite all enthusiasm for internal markets, managers are obviously not vanishing off the face of the earth. Instead, a complex suspension of power relationships appears to be taking place amongst those involved through the strengthening of market principles.

3.1 The fiction of the pure market – the meaning of the social embedding of markets

Neo-classical market theory and economic theory take as a starting point the fact that various contractors offer their goods and services on a market and that those both interested and financially capable compare the quality and price of the services and goods on offer with one another. The coordination takes place in the markets exclusively on the basis of price, from which all relevant information about quality, deadlines and delivery capabilities is calculated. This leads to a contract between the buyer and seller in which service and consideration are thoroughly specified (cp. the discussion in Baker 1990, p.591).

This ideal typical market seems fair because cooperative relationships for longer periods are not catered for in market transactions. In the market transaction, the vendor is only interested in the buyer's ability (or inability) to pay, and not in his political leaning, gender, nationality or religion. From this perspective, markets appear to be quasi- democratic models with the motto "one euro equals one vote" which distances itself from consideration of "social standing and class", "morals and religion" or "family and friendships".

In this line of thinking, dynamic sampling of the organisation through principles of marketisation and competition can be considered. It is argued that in these companies the capability, motivation and creativity of employees is no longer inhibited by managers who came into their position years previously and now belong to an "old boys" network in which they help one another further up the ladder. Instead, a climate now predominates in which everybody, irrespective of their gender, origin or skin colour, can assert themselves on the internal market by way of their own achievements.

In practice, however, markets have very little in common with this ideal type. Even the grand seigneur of French sociology, Émile Durkheim, developed the idea that only the "non-contractual part of the contract", such as trust and social relationships, enable anything like a free market to develop. From this perspective, markets are not "natural phenomena" which occur when social development is left well alone, but are rather the result of social construction (see Friedberg 1993, p.9).[6]

---

[6] From the perspective of society as a whole, Polanyi (1994) developed the idea that the implementation of market principles is always conducted by organised regulatory measures. From this perspective, markets do not develop through liberation of the economy from state control, but on the contrary are made possible *because* of the state as a central authority. Generally accepted formalities such as property rights, freedom of contract and legal security repress free market development and thereby make markets at all possible.

Several empirical studies have since demonstrated the way in which market transactions are embedded in a multitude of social processes. It has been shown that even for supposed prototypes of markets, such as national securities markets, stock markets or agricultural auction markets, the social relationship between traders strongly influences the transaction. The existence of money as a means of exchange has itself neither made direct negotiations redundant nor depoliticised the exchange (cp. Moullet 1983; Baker 1990).

In light of the social embedding of markets, scholars speak of a "fiction" or "social myth" of the pure market and demand that markets be observed in their actual form (White/ Eccles 1986, p.135; Friedberg 1993, p.128). The question therefore begs to be asked how internal markets in which the "intrapreneurs", "one-man-enterprises" and "self-Ltds." operate are structured.

3.2 The meaning of objectives, membership and hierarchies in organisations and the important differences between internal and external markets

Directly because markets are socially embedded, it is necessary to bring out the differences between external and internal markets. The boundaries of a company mean that internal market relationships can be organised very differently to the market relationships between the business and its suppliers, clients or partner firms. The assumption that markets are socially constructed is to be understood as a demand for greater differentiation rather than obliteration of the boundaries between the organisation and its surrounding environment. Using three features- objectives, membership and hierarchies- it can be shown that whether a market exists in a social subsidiary system of the economy or in an organisation is a central difference.

On objectives: in contrast to the societies of both antiquity and the middle ages, modern societies do not commit themselves to high-ranking objectives such as the happiness of the people, racial purity or the fulfilment of God's will.[7] Accordingly markets, as an important method of coordination within the economy, are not subordinate to a superior objective. Complaints about the "terror of the economy" refer to the fact that markets de-

[7] Societies which nowadays *do* commit themselves to such an objective risk being discriminated against as totalitarian, ideological and outdated.

velop a life of their own and elude every form of superior “humanitarian” objectives. In contrast to markets: irrespective of whether in an administrative department, a high-tech company or a trade union, objectives such as a more or less amicable gratification of enquiries into identity cards, the penetration of a new ultra-light mobile phone onto the market or the conclusion of a trade agreement with large pay increases play a central role in the orientation of organisations.

From this central distinction between organisations and objectives, it is clear that internal markets are much more firmly orientated towards objectives than external markets. Business divisions, profit centres or departments in internal markets cannot assume that being in the black is a guarantee for continuance in the company and therefore in the internal market. Outstanding performance on the part of an intrapreneur or a one-man-enterprise is just as small a guarantee of their continuance within the corporate group. Urged by strategic advisors to do so, companies constantly change their core competences and redefine their objectives. When a participant in the internal market is at the current time no longer well suited to the negotiated objectives, he is excluded from the internal market processes, quite irrespective of how he is performing. The motto within a company is not simply “do what you want but be profitable with it” but “do what you want but be profitable with it and stick to the (ever-changing) objectives of the corporation.”

But the reverse process is also to be seen. A profit centre can continue to exist within a business, even when it has only been making losses over the past few years. While a company with such a deficit would quickly go bankrupt in the “free market economy”, it can continue to exist within the internal market because it is financed by profitable profit centres within the system. The background to this development is that the enterpriser does not only work with the aim of being profitable. Motives such as personal fulfilment and Christian missionary work can also play a role.

On membership: A total exclusion from society only takes place in exceptional cases. Most modern states to not revoke deviant citizens′ nationality. As the most radical form of exclusion, the death penalty is only used in very few “civilised states”. Even the “free” market refrains from global exclusion of buyers and sellers. A machinist would have difficulty in justifying himself if he did not sell his machines to the highest bidder. A buyer of telephone systems does not exclude vendors from the market process simply as a matter of principle. Everybody can participate in external markets, provided they offer appropriate payment.

In contrast to markets, the management of membership is a central feature of organisations. Who belongs to an organisation and who does not is selectively determined (Luhmann 1995, p.16) and this has implications for internal markets. Even though it is suggested in management literature that the "enterpriser within the enterprise" makes his own decisions about what is marketable, decisions about continuance or discontinuance are made within the organisation. A profit centre does not go bankrupt but is always either closed or sold. An intrapreneur similarly does not go bankrupt but is dismissed in the usual manner.

This difference between "inside" and "outside" makes it possible not for market processes to invade unfiltered into various units but on the contrary to work in companies with simulated markets. Internal markets are structured in such a way that the achievements of a profit centre or intrapreneur can be compared to market prices on "free" markets, but as a rule the members of a business are granted the opportunity to accept the lowest price of external vendors. The settlement prices between individual enterprisers within the business are not negotiated by a free-for-all contest of abilities but are determined by company headquarters.

On hierarchies: the age in which society was organised by a strict hierarchy is long since gone. There is no longer any king, emperor or pope who, through his chain of command, can govern various spheres of the lives of the people. Nobody would accept George Bush or Romano Prodi as a hieratically superior leader, the only exception being the members of the White House or the officers working for the European Union. In contrast to modern societies, hierarchies are a central feature of organisations. Despite all the rhetoric about decentralisation and abolition of hierarchy, organisations are unimaginable without hierarchy. It is because of this hierarchy that unions, administrative departments and enterprises can act at all like computable, collective entities because through direction, assurances from management level of the organisation can be fulfilled (Luhmann 1997, p.834).

Internal markets are not at odds with hierarchical coordination in the enterprise, but on the contrary are included in the hierarchical structures. It is hierarchical decisions which decisively structure internal market mechanisms. The recruitment of employees, their assignment to positions, their dismissal or promotion and their salaries form the basis of organisational criteria which are decisively determined by the hieratical head of the company. They are not in some way dictated by anonymous market mechanisms (Scott 1986, p.255).

It cannot be inferred from the distinction between internal and external markets by means of objectives, membership and hierarchies that a great bureaucratic conspiracy on the part of companies′ top management lies behind the marketisation of internal processes. It is often overlooked that decisions about objectives, membership and hierarchies are not centralised at the head of the company. A company's objectives can change without any strategic decisions from top management. An employee can resign from a company when an offer from another firm is more appealing. Even hierarchies themselves can erode without any interference from management level. But one thing is clear- employees do not become the new leaders of the business as a result of the conception of "intrapreneurs", "one-man-enterprises" or "self-Ltds.".

4 Myth: the concept of intrapreneurship promotes the integration of employees into the business

It is complained about in management literature that employees were never formerly perceived as important "human capital" and as people were kept out of organisation as much as possible. Employees were not allocated a central role in company rhetoric until the seventies and eighties. "People are the most important resource in the company" (Deal/ Campbell 1982, p.15). "Excellent companies consider their employees as a source of quality and productivity enhancement" (Peters/ Waterman 1983, p.37).[8]

Only through the use of management concepts such as "intrapreneurs", "one-man-enterprises" or "self-Ltds." is it possible to integrate employees and their creativity, productivity and engagement into a company. Employees would have more job satisfaction because of extended freedoms and would therefore identify more strongly with the company.

These notions of the integration of employees by means of the intrapreneurship concept is at first surprising because one would sooner expect a gaining of independence within the company followed by a willing inte-

[8] Without having to question the claim to originality of Deal/ Campbell and Peters/ Waterman, the ideas can be traced far back. The basic idea can be found in Karl Marx, who stipulated that the capitalist can only make a profit by exploiting external workers.

gration in an organised context. An enterpriser is no longer an "organization man" (Whyte 1958) or "corporate" (Kanter 1983) but is a trader in his own affairs. How is it, then, with the integration of the intrapreneur?

4.1 The dilemma of simultaneous integration and exclusion of employees

The idea has in the past been developed, in the example of the classic bureaucratic organisation, that companies' management faces the dilemma of integrating their employees into the firm whilst at the same time keeping the option of excluding them open. Within enterprise there is a contradictory demand of both integrating employees, so that their creativity and engagement can be utilised, and also maintaining the fact that they are replaceable, in order not to become dependent on them.

It is argued in system theory that, in contrast to families, organisations are based on the principle of exchangeability of people. Organisations are comprised of members who are only partly integrated. In modern society, one does not become an employee of a company to the exclusion of all else. This only partial integration of employees on the one hand relieves the burden on individual employees because a dismissal from a company does not simultaneously mean an exclusion from other organisations such as sports clubs or even society as a whole.[9] On the other hand however, this also relieves the burden on organisations because they must no longer feel responsible for their employees as people. The unscrupulousness with which employees are fired is only possible because heads of personnel can be certain that a dismissal does not generally mean an exclusion from society as a whole.

How integration can be organised in enterprises without the destinies of the company and the employee becoming intertwined is the result of everyday processes of negotiation. On the one hand, organisations are constructed in such a way that managers can be replaced as well as employees without the organisation falling apart as a result. These expectations of employees´ conduct are determined through hierarchical positions of rank and by programmes. They specify who should talk with whom and how. However, at the same time company processes are not programmed in such a way that an organisation can operate like a machine. The organisa-

[9] It is because of this that mass unemployment is so problematic because an exclusion from an organisation is increasingly felt to be an exclusion from society and therefore the pressure on both employees and companies increases.

tion is dependent on the willingness and capability of its individual employees to deal with everyday fluctuation of the company's objectives (Luhmann 1995).[10]

There is no eternal compromise for the relationship between employees´ inclusion and exclusion. It is always negotiated afresh. How has this been solved traditionally? How does this relationship alter when an enterprise tries to make its employees "enterprisers within the enterprise" using the concepts of the "intrapreneur", "one-man-enterprise" or "self-Ltd."?

4.2 The paradox of an "enterpriser within the enterprise": the enhancement of employment contracts with elements of contracts for work and labour

The contradiction of inclusion and exclusion of employees was classically connected to the employment contract. Although employers like to complain about what they are letting themselves in for in regard to obligations, worries and problems, it cannot be overlooked that enterprises, administrative departments and unions attain a high level of flexibility through employment contracts. While service and return service are specified exactly in a sales contract, such as in the purchase of a stamp or training package, with an employment contract the employer purchases labour in a very abstract form. By signing a contract of employment, the employee is effectively writing out a blank cheque and declares himself prepared to utilise his own manpower, capability and creativity for the purpose of the tasks presented to him. He does not have written into his contract in which areas his performance will lie (cp. Commons 1924, p.184).

[10] Cornelius Castoriadis (1997, p.137 et seq.) described the same process from a Marxist perspective. He argues that there is a fundamental contradiction in capitalism in the fact that employees are simultaneously treated as objects of bureaucratic manipulation and also as self-acting subjects. Management is instructed to exclude employees from production processes in order to maintain the manageability of such processes but at the same time is also instructed to keep employees involved in the process because the necessary flexibility of the production process can only be guaranteed using knowledge acquired in situ. Castoriadis alludes to this contradiction, that processes of creation of value can never run as trivial machines can, but rather always require flexible adjustment from employees who possess the production "know-how". This "know-how" also forces employees to guarantee the functioning of the system by going against the system´s own rules to do so, i.e. to deviate from the official rule book in order to ultimately guarantee production.

The deal between employer and employee is structured in such a way that the employee submits to the company objectives, promises to obey hierarchical instructions and is rewarded by the employer with payment, share options and/ or career prospects (cp. Barnard 1938, p.167 et seqq.). A "zone of indifference" is generated, along with its many consequences, inside of which employees may not refuse to follow the orders, demands, instructions and guidelines of their superiors (Simon 1957).

The advantage to company management is clear: employees promise a form of general obedience to orders and instructions specified no further. They therefore enable management to adapt the organisation to altered demands very quickly and without complicated internal processes of negotiation. If every member were allowed the right to permanently offer his opinion to discussion of the structure of the organisation, the formation of flexible, complex structures would never take place (Luhmann 1995).[11]

The company management pays for this advantage of flexibility, however, with a control and integration problem: because the employment contract does not specify the exact nature of the employee's achievements, the employee could try to limit his performance. Unlike a contract for work and labour, in which performance is specified exactly and in which it is in the contractor's interest to produce results as quickly as possible, it is implied that employees try to conserve their labour. Management reacts to this threatening withdrawal of labour through control. From this perspective, the whole history of enterprises in modern industrial society can be interpreted as a struggle for control of this labour.[12]

When companies now experiment with the concept of the "intrapreneur", the "one-man-enterprise" or the "self-Ltd.", something significant has occurred in the relationship between inclusion and exclusion. The replacement of employees with independent enterprisers or the substitution of employment contracts with a variety of contracts of work and labour are clearly not taking place in enterprise. "Independent sub-enterprisers", "quasi-autonomous workers" and "franchisees" are increasing as a general model but organisations do not appear to be solely dependent on this

[11] A whole branch of organisation theory is based solely on the idea that the efficiency of a company can be directly put down to the use of the employment contract to enable action without time-consuming processes of communication (cp. as a starting point Coase 1937).

[12] And this can also be said of a theoretically demanding and empirically interesting school of thought in industrial sociology- the Labour Process Debate.

model. A development based only on independent sub-enterprisers would be a waste of the organisation's potential for flexibility and efficiency: the advantages of flexibility attained by the organisations through employment contracts would be lost. Performance would have to be specified exactly in every completed contract and it would be very expensive for every operation to be financially appraised and compared with offers from other contractors (Coase 1937).

In contrast to "real enterprisers"- and this is an important distinction- "enterprisers within the enterprise" do not become the proprietors of the means of production relevant to their creation of value. The unity of "having" (the means of production) and "doing" (the yielding of results), as is the classic case with the self-employed, does not exist for the "intrapreneurs", "one-man-enterprises" and "self-Ltds.". The formulation of the "enterprisers within the enterprise" therefore seems paradoxical.

In the case of the "intrapreneurs", "one-man-enterprisers" and "self-Ltds.", the employment contract is enhanced with elements of the contract of work and labour.[13] A central instrument with which the employment contract can be combined with elements of the contract of work and labour is target agreement. Through the use of this mechanism, market pressure can be produced within the organisation in two ways: firstly, employees are connected to the market conditions of the company with demands such as "10% profit margin" or "5 % more new clients" (negotiated). Secondly, target agreements can establish internal competition.

4.3 The internalisation of the exclusion and inclusion dilemma

Everything indicates that the dilemma of exclusion and inclusion is not being resolved in the new work structure but is merely being relocated. To both exaggerate and simplify: in the classic tayloristically organised firm, the sides of the conflict between inclusion and exclusion were clear. On the one side was the management, who tried to make as much use as possible of the abstract manpower of employees, already purchased in a lump sum, without becoming involved in a one-sided relationship of dependence. On the other side were the employees, with an interest in conserving their own manpower but at the same time wanting to be as indispensable to

[13] This model has existed previously- consider for example sales representatives. The difference is that an attempt is now being made to apply it throughout the whole organisation.

the company as possible in order to move their own market value up the ladder.

To again exaggerate, the concepts of intrapreneurs, one-man-enterprises and self-Ltds. tend to shift this conflict to the individual employees. Oswald Neuberger (2000: 73) argues, following Michel Foucault's line of thought, that control no longer comes from one omnipresent boss but that the market is elevated to a single boss from whom nothing escapes and who objectively rewards success and harshly punishes mistakes. The impression develops amongst employees that failures as intrapreneurs are punished not by their superiors but by the alleged "objective consequences" of their own actions. Dismissals and closures of departments are no longer perceived as the arbitrariness of an enterpriser geared towards maximising profits but as the logical consequence of the internal market.

With the introduction of entrepreneurship into the company, every employee is now himself confronted with the problem of inclusion and exclusion. Klaus Peters described the process in such a way that the two wants, which in the bureaucratic, system were neatly allocated to two different people- here the employee who really just wants to go home, and there the boss who wants to detain him against his will- now simultaneously coincide in the same person due to the concepts of intrapreneurship. On the one hand the intrapreneur does not want to work any longer than he has to but on the other hand wants to return to his desk (Peters 1999, p.8).

Questions which were in the past classically posed by top management suddenly emerge as questions of the intrapreneur: is my contribution part of the company's core competence_or should I develop my capabilities differently? Is my contribution enough for the company or must I offer more? Does my labour really have a current market value or am I merely a burden for the firm?

5 Conclusion: paradoxical behavioural demands in marketised companies

It would be all too easy to describe the concepts of the "intrapreneur", "one-man-enterprise" or "self-Ltd." as merely the latest publicity gimmick with which to guide employees. Certainly, these management concepts are not implemented into the company on a one-to-one basis. They suggest a conclusiveness which in practice is not to be found.[14] With the concepts of the "intrapreneur", "one-man-enterprise" or "self-Ltd.", a development, already foretold in the decentralisation craze of the nineties, is being radicalised: the contradictory environmental demands which face a company are only intercepted by organisational structures to a limited extent, and are more commonly passed on as behavioural demands for employees to deal with instead.

5.1 From the unambiguousness to the inconsistency of behavioural demands

The sociologist James D. Thompson (1967) pointed out that for a long time a central strategy for companies, administrative departments and unions has been to seal off a company's core for creation of value from uncertainties or contradictory demands. Through the development of special departments to deal with uncertainties such as operations scheduling, personnel and organisation, or buying and selling, the assembly line of a car manufacturer, the warehouse and shipping department of a wholesale company or the processing department of an administration are greatly secluded from the uncertainty of the environment and are provided with largely unvarying information.

The division between departments dealing with uncertainties on the one side and a largely stable core for creation of value on the other would in many cases lead to monotonous and tedious working conditions for employees working at the conveyor belt or in the shipping or processing departments. There would also, however, be advantages for these employees. They could expect their job instructions to be formulated in such a way as to be achievable. Wilfred Glißmann rightly called attention to the fact that the exercising of power must involve instructions or orders which are in principle accomplishable. According to Glißmann, severity and cruelty cannot be ruled out and the fulfilment of an instruction or order can result in severe injuries. In the most extreme case imaginable, a command is only accomplishable if the employee who carries it out loses his life. But the following is also true: orders must be accomplishable. Unrealisable commands would damage the legitimacy of the person giving out the orders and the system of command as a whole. If the person giving out orders ut-

[14] Approaches centred around employees do not therefore differ fundamentally from other management concepts which are similarly not to be found in their pure form.

ters an unrealisable command he will no longer be taken seriously by his colleagues (Glißmann 2000:9).

With the concepts of "intrapreneurship", "one-man-enterprises" or "self-Ltds.", this "protection" is removed. The functioning of secure roles, to which Luhmann draws attention, tends to be retracted. Employees no longer know what to expect, what they may do and what they may not. There is no longer any protection from the mood swings of those with power within the organisation, nor any relief from limitless responsibility.

5.2 Paradoxical behavioural demands

Employees understand this reorganisation as paradoxical behavioural demands. On the one hand, intrapreneurs are required to compete with other intrapreneurs. On the other however, they should be able to cooperate with other employees. Motto: everyone is pulling in the same direction but only the best prevail. On the one hand, intrapreneurs are required to go their own way, but on the other should not lose sight of the common objectives of the company. Motto: everybody seeks their own way but we all sit in one boat. On the one hand, enterprisers within the enterprise should violate rules and regulations, but on the other should respect the structures set out by the company. Motto: do what you want but do not violate the written and unwritten rules.

Because of these paradoxical behavioural demands, employees find themselves in a classic double-bind situation, i.e. they are confronted with contradictory behavioural demands, both of which must be simultaneously carried out. When only one of the behavioural demands is carried out, the other is being disobeyed. When the employee switches here and there between the behavioural demands, he runs the risk of being accused of unreliability and of having no clear policy.

Oswald Neuberger (1991: 46) pointed out that a possible way of avoiding these paradoxical behavioural demands is to begin a meta communication of them. Employees should internalise the paradoxes of "dependent autonomous workers" and "entrepreneurial behaviour in the enterprise" not as their own problems but as a contradiction inherent in the construction of this new form of organisation. The consequent demythologisation of the concepts of the intrapreneur, one-man-enterprise and self-Ltd. is merely the first step towards this. To adapt an assertion of Rainer Werner Fassbinder (and ultimately of Karl Marx): when one cannot immediately change something due to its paradoxical character, one should at least describe this paradoxical character (see also Kühl 2002).

Translated by Alice Ashworth

References

Baker, W. E. (1990), Market Networks and Corporate Behavior; in: American Journal of Sociology, Jg. 96, S. 589-625.

Barnard, C. I. (1938), The Functions of the Executive, Cambridge, MA.

Castoriadis, C. (1997), Gesellschaft als imaginäre Institution. Entwurf einer politischen Philosophie, 2. Aufl., Frankfurt a.M.

Coase, R. H. (1937), The Nature of the Firm, in: Economica, Jg. 17, S. 386-405.

Commons, J. R. (1924), Legal Foundation of Capitalism. New York.

Deal, T.E.; J. Campbell (1982), Corporate Cultures. The Rites and Rituales of Corporate Life, Reading.

Friedberg, E. (1993), Le pouvoir et la règle. Dynamiques de l´action organisée. Paris.

Fröhlich, D.; U. Pekruhl (1996), Direct Participation and Organisational Change. Fashionable but Misunderstood? An Analysis of Recent Research in Europe, Japan and the USA, Dublin.

Glißmann, W. (2000), Ökonomisierung der "Ressource Ich" - Die Instrumentalisierung des Denkens in der neuen Arbeitsorganisation; in: Denkanstösse. IG Metaller in der IBM, H. Mai/2000, S. 5-24.

Gutenberg, E. (1983), Grundlagen der Betriebswirtschaftslehre. Die Produktion, 24. Aufl., Berlin.

Kanter, R. M. (1983), The Change Master: Innovation for Productivity in the American Corporation. New York.

Luhmann, N. (1995), Funktionen und Folgen formaler Organisation, 4. Aufl., Berlin.

Luhmann, N. (1997), Die Gesellschaft der Gesellschaft, Frankfurt a.M.

Moullet, M. (1983), La Concurrence organisée, Paris: Doktorarbeit am IEP Paris.

Neuberger, O. (1991), Personalentwicklung, Stuttgart.

Neuberger, O. (2000), Das 360 Grad-Feedback. Alles fragen? Alles sehen? Alles sagen? München; Mering.

Peters, K. (1999), Woher weiß ich, was ich selber will? Die Abschaffung der Stempeluhr bei IBM und die Frage nach den Interessen der Arbeitsnehmer; in: Denkanstösse. IG Metaller in der IBM. H. Februar/1999, S. 3-10.

Peters, T. J.; R. H. Waterman (1983), Auf der Suche nach Spitzenleistungen. Was man von den bestgeführten US- Unternehmen lernen kann, Landsberg am Lech.

Polanyi, K. (1994), The Great Transformation. Politische und ökonomische Ursprünge von Gesellschaften und Wirtschaftssystemen, Wien.

Schmidt, A. (2000), Mit Haut und Haaren - Die Instrumentalisierung der Gefühle in der neuen Arbeitsorganisation; in: Denkanstösse. IG Metaller in der IBM, H. Mai/2000, S. 25-42.

Schumpeter, J. (1926), Theorie der wirtschaftlichen Entwicklung. Eine Untersuchung über Unternehmensgewinn, Kapital, Kredit, Zins und den Konjunkturzyklus, 2. neubearb. Aufl., München.

Scott, W. R. (1986), Grundlagen der Organisationstheorie, Frankfurt a.M.; New York.

Simon, H.A. (1957), Models of Man. Social and Rational, New York.

Smith, K.K.; D.N. Berg (1987), Paradoxes of Group Life, San Francisco.

Thompson, J.D. (1967), Organizations in Action, New York et al.

Warnecke, H.-J. (1992a), Die Fraktale Fabrik. Revolution der Unternehmenskultur, Berlin et al.

White, H. C.; R. G. Eccles (1986), Control via Concentration; in: Sociological Forum, Jg. 1, S. 131-157.

Whyte, W. F. (1958), Lohn und Leistung, Köln.

“Performance Community” – New Financing Solutions

Anna-Maria Schäfer

LBBW Landesbank Baden-Württemberg, CF Focus Industry Automotive

Automobile manufacturers (OEM) and automobile suppliers are increasingly in a way connected with one another, which deviates from a conventional and simple buyer-vendor relationship. Both partners are connected by agreements, obligations, rules and contracts in a way which often affects very profoundly the business models and processes of each other. Above all the transfer of value-added operations or tasks (e.g. engineering, R&D, logistics, distribution....) of the OEM towards the suppliers brings them more and more into an area of conflicts between internal and external factors. Their own processes, structures and strategies are affected accordingly or must be modified. At the same time also a transfer of risks and problems takes place. Additionally the financial situation is possibly not as healthy as it should be to handle large orders or projects. Traditional or classical fund raising methods may not cover all aspects of this specific situation. A closer look to alternative ways of financing in consideration of new partnering models is requested.

Powerful partners are required

First are to be stated that the car manufacturers itself struggle in different specifications with own internal restructuring, optimization and cost problems. This is not without effects on subordinate enterprises as well. Above all, the levers cost reduction and price reduction in functions such as development or purchasing provide for tensions in the co-operation with the suppliers. The OEM has naturally a fundamental interest in an efficient and economically healthy supplier who is able to comply with the demanded achievements. Also the bank as a possible financing capital provider on the other side has - with some different perspectives - the same basic interests as the OEM. From these so far respectively bilateral collaborations the future will come up with a new defined co-work between the OEM, the supplier and the financing partner. The result will be a trilateral relationship or partnership.

Shifting of responsibilities

The current trend shows that on the one hand the OEM are transferring increasingly achievements or tasks towards their suppliers, on the other hand however doesn't want to take over all or part of the arising costs or financing expenditures respectively. At the same time the tasks and/or orders are even more challenging, more complex and targeted to a long-term relationship. Thus develop also hardly reversible dependence and connections especially in the case of already started projects or very specific tasks.

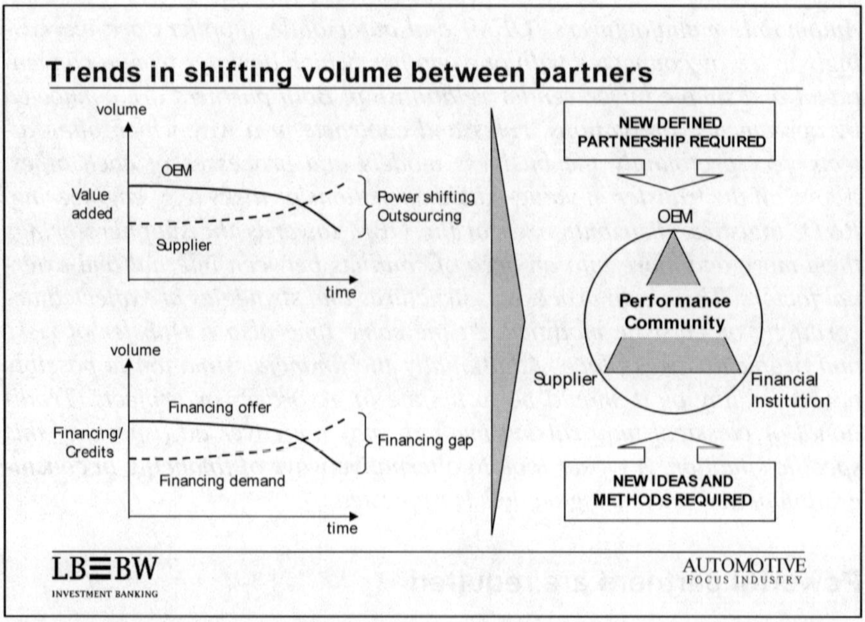

Fig.1: Shifting volume

Engineering projects may reflect the following situation in particular: The supplier receives the order to develop certain new components or systems/modules for a new car model range. This implies normally a 2-3 years time span for conception, pre-development and production stage conversion of the assigned components. At the same time all the necessary tools, machines or processes have to be prepared. The supplier must be able to supply the developed components in the ordered number of items from the vehicle's start of production (SOP) over a certain runtime period to the OEM. During this described development and preparation phase an amount of investments originates, which the supplier must calculate and plan with his resources. The OEM refunds a certain partial amount, it remains however a remainder, which has to be generated by the supplier in own contribution. The amortization of these pre-investments is based upon the agreed unit price for the developed components. Only then, if the components are going into the vehicle serial production, the appropriate amortization runs as contracted.

This constellation may cause a gap in the suppliers financing potentiality and requires additional efforts. More often the premises for this situation are not ideal: the company has too little own capital or cash funds or is already bound to other financing strains and can cope with additional expenditures only with difficulties. Additional debt financing deteriorates the companies balance structure. If the bank is then asked for a financing solution, the financing mechanisms and approaches are requiring a detailed analysis of the enterprise (annual accounts) and the given projects. The calculation of a loan for a company will be based upon new defined rating criteria (Basel II). In the future the borrower receives a rating classification which reflects the solvency. Balance sheet relations are of importance. A low rating level and a negative risk structure are resulting in increased capital costs. For notoriously equity-weak companies with a low solvency

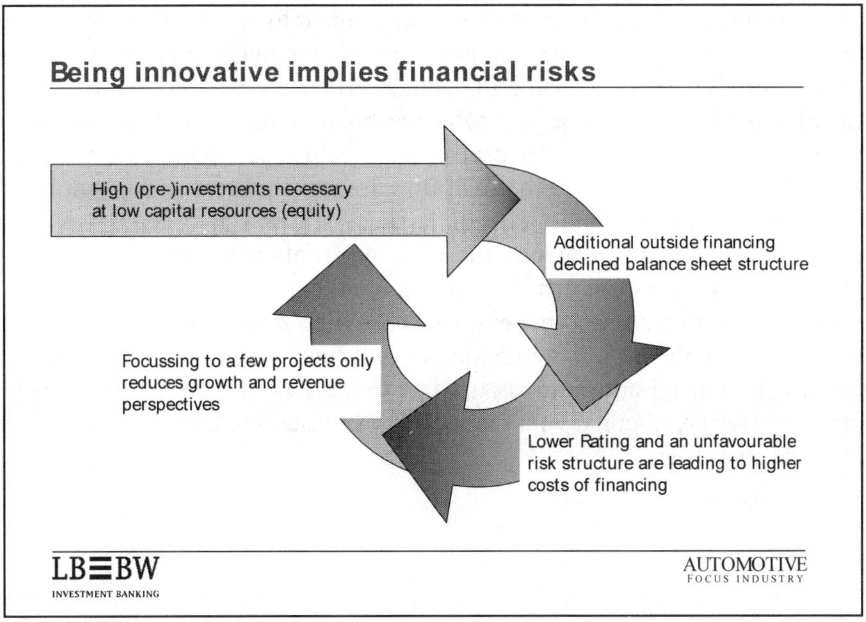

Fig.2: Financial risks

level this may be momentous: Only a reduced number of projects or shortened project volumes can be handled under the given circumstances. This fact in turn reduces the actual growth and earnings perspectives of the company additionally.

Alternative methods and concepts

Based on the described starting position it is necessary to think about suitable instruments and methods which try to combine the interests and possibilities of all involved partners. A "Performance Community" replaces traditional viewpoints or opens new dimensions. Important is that the partners co-operate in a trilateral relationship and pursue common interests. Especially for the supplier, who has to finance future projects, innovative and professional financing alternatives offered by the financing partner are of importance. The classical bank loan deserves its authorization as commonly used and accepted method, it can however not consider all aspects of the illustrated challenges regarding an engineering project. It is mainly focused on past-oriented figures as well as securities (properties) referring to the entire company. From a legal and fiscal point of view there is only an on-balance representation possible. In contrast to the classic loan the alternative financing tools like the object financing and/or project financing (see illustration) can be related to a single project only. This enables a so-called off-balance solution, i.e. balance unburdening. Important here: the base for the evaluation of the project is its future economic success. The financiers look primarily to the cash flow from the project as the source of funds and its assets. Future economic success means also: the developed component will be integrated in the vehicles serial production and the supplier receives a pay back per delivered unit. Securities are the tools or machines to produce this component, because without this tools it is impossible to produce the required components and finally to produce the vehicle itself. The crucial advantage is the fact that the project-dependent development costs are coupled with the tooling expenses and both parts are used as credit securities.

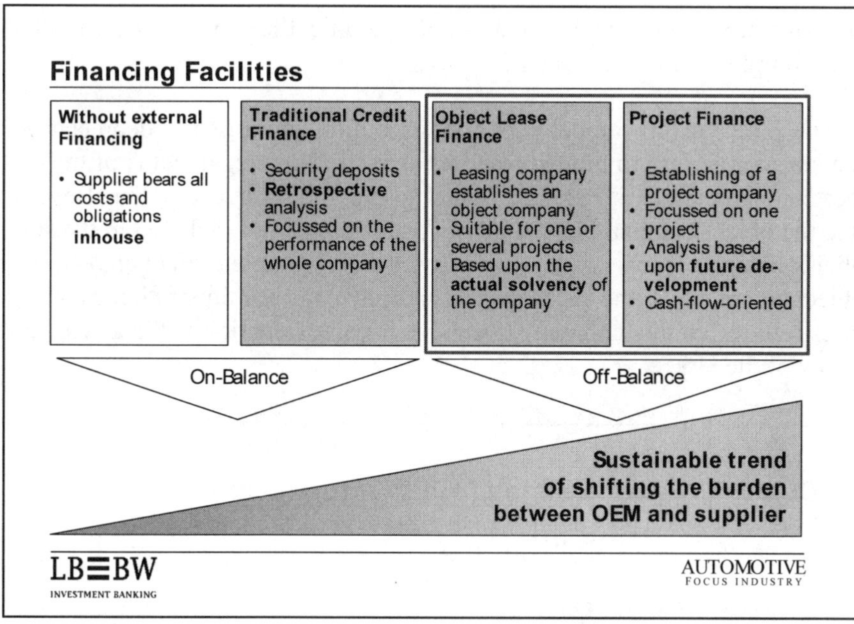

Fig.3: Financing Facilities

Off balance solutions

Below outlined are the two alternative financing methods "object financing" and "project financing".

Object financing (Joint-Venture Leasing)

Structure: A leasing company provides an object company, this object company covers the pre-financing of engineering or development work and the necessary tools or machinery for one or several projects (here "object"). The results of the project work (e.g. ability to supply the developed automotive components) are leased back to the automotive supplier (operating-lease-contract). Prepayments to secure the ongoing engineering work have to be made by the object company. Calculations consider an open residual value for the tools/machinery due to balance neutrality. Repayment obligations are linked to the sale of the produced components. An additional basic contract premise is the commitment of the supplier to deliver the demanded components to his customer (e.g. OEM). The duration of the leasing contract is shorter than the duration of the supply agreement. In case of a contract cancellation ahead of schedule, the supplier is committed to accomplish a full repayment.

The joint venture model uses the tax optimization of the depreciation, the constant access to hidden reserves as well as the legal status of the object company, since it belongs factual to the automotive supplier. Beyond that the object company is treated off-balance, it is not included in the supplier's balance sheet. All leasing activities are completely covered by the object company, so the supplier does not have any administration expenditures referred to this company. Decisive for an evaluation is the actual solvency of the lessee.

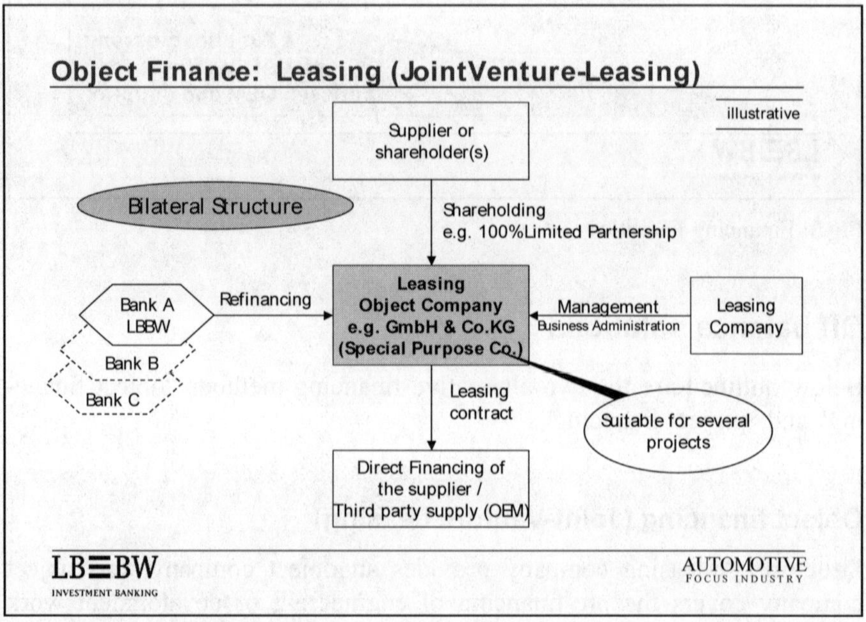

Fig.4: Object finance: leasing

Project financing

The main focus of a project finance model is directed towards the project company which is a so-called "special purpose company". The assignment to develop new components or systems/modules given originally to the automotive supplier will be shifted to this project company. It is provided with all the necessary properties (tools, R&D results and output, legal

rights, patents) to assure the delivery of the requested components. In fact the supplier develops and produces the components for the project company based upon an operation management agreement. The supplier is also liable for the proper delivery/production of the components. The project company is the borrower of the funds to finance the specific project. In case of the project company the bank acts as general partner with a minimum of 51 % of the voting rights and normally also more than 50 % of the capital shares. With this described constellation the off-balance structure is applicable as there is no consolidation of the project company within the accounts of the supplier. The essential element of a project financing is the look at a real project-related payment flow. Again the idea is to look at the future performance of a project, as repayments are calculated and based upon the number of units produced for the serial production of the vehicle. Securities are depending from the future cash flow related to the project.
A basic element of such a financing model is the look at the possible risks. It is necessary to analyze the details of the given project: What kind of component is demanded, is it a new development or a further development of an already existing part. Other aspects like technology, market, competition, pricing, importance, model range, serial or optional equipment etc. have to be included and evaluated as well.

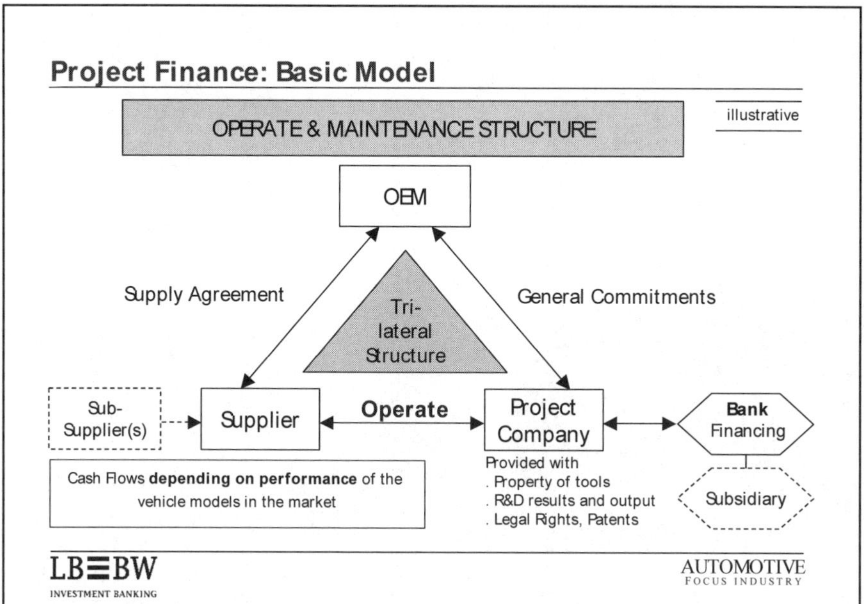

Fig.5: Project finance

Perspectives

Changing relationships between the business partners are leading to modified approaches in terms of financing methods or options. The bank responds to the changes with new tailored financing models. But it has to be pointed out, that the complexity and dimension of realizing a project with these new opportunities is not a simple venture. Project content, definitions, terms and conditions have to be analyzed and evaluated not only with financial performance measurements but also with industry-related factors. This includes also a learning curve or target for all parties involved. Chances and risks have to be balanced. The decision to use these instruments is no longer a stand-alone decision; it reflects a new defined partnership between OEM, supplier and bank.

And finally not all kind of projects is suitable for such "treatment". A minimum size (project volume) of about 5 million € tags the level for an object finance solution, for a project finance solution the base level should be about 20 million €.

Identification of Cost-Saving Potentials

Achieved by Collaboration Using Product Lifecycle Costing

Alexander Höppe

IBM Business Consulting Services

The cost-saving potentials of collaboration with partners, suppliers and customers over the entire product life cycle are enormous. However, a reliable identification of these saving potentials requires a cost-orientated view of the product life cycle, i.e. a "product lifecycle costing". Despite economic slumps, multiple project experience shows that companies from all industries take very little advantage of cost reduction achievable through collaboration, because existing IT-systems lack collaborative product lifecycle management functionalities. Focusing on engineer-to-order companies (ETO), this article provides information, in particular, about cost-saving potentials brought about by collaboration over the complete product lifecycle. Furthermore, it describes how these saving potentials can be identified if an effective product lifecycle costing takes place.

Product Lifecycle Costing– Definitions and Explanations

When searching for saving potentials, companies have to deal time and time again with questions such as "What do our products actually cost?" To find an appropriate answer to this question, it is essential to evaluate the product with regard to its costs, from the first product idea right through to its recycling stage. These costs are called the lifecycle costs (LCC's) of the product. Using the ETO example, Fig.1 demonstrates the relevant costs over the whole product lifecycle. Furthermore, the approximate time of occurrence can be seen.

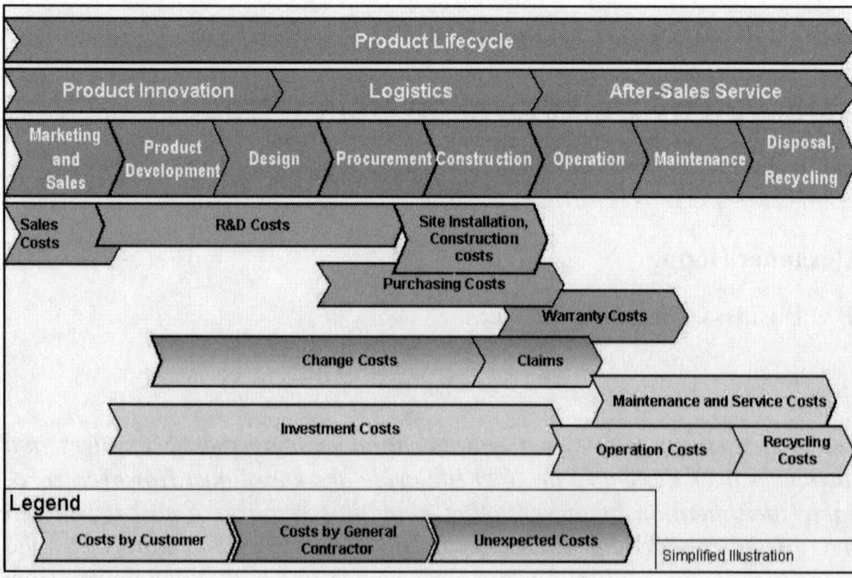

Fig.1: Product Lifecycle Costs (LCC's) using the example Engineer-to-Order (ETO)

Product lifecycle costing means taking a cost-oriented look at the lifecycle. In doing so, the identification of cost drivers and the controlling activities in order to maintain the target costs (primarily cost of goods manufactured) are especially important.

A cost target is a quantitative measure for manufacturing, operational or maintenance costs. This value is determined by market forces and consists of different cost elements, e.g. cost of goods manufactured (supplier side) or operational and maintenance cost (final consumer / operator side).

In the case of engineer-to-order (ETO), product cost controlling and project controlling are closely related and can both be considered synonymous to LCC-controlling. Due to the fact that within ETO customised and unique products are involved, the product structures are (possibly automatically) carried forward later on in the ERP system as project work breakdown structures (WBS). The project WBS within ERP makes it possible to combine the product lifecycle phases and the product structure, so that the wide area of project management and cost analysis functionalities of the ERP system can be applied to the product.

Current Situation

But how can product lifecycle costs be calculated? Is it possible to determine planned costs and budgets for newly developed products with an acceptable degree of accuracy? Does the system provide reports to monitor product costs at any time and at any stage of the product lifecycle? The answer to these questions is difficult for the following reasons:

- Product costs are not usually available until the project is finished.
- Cause and time of cost accruement are difficult to track and determine.
- A lack of system communication between product design and after-sales service makes a determination of total product life-cycle costs almost impossible. Informed decisions, which are essential to a successful project management, cannot be made.
- Delayed cost controlling caused by inhomogeneous IT-infrastructures, non-existing or weak definitions of responsibilities and insufficient communication.
- Missing collaboration between product development partners, lack of coordination in the case of collaborative project and programme management activities due to organisational weaknesses and low level IT-integration or non-existence of appropriate system functionalities.

Without clear information regarding product lifecycle costs, a cost-orientated control of projects is not possible, resulting in incorrect or insufficiently investigated decisions, the financial impact of which is unpredictable. This in turn does not generate any learning effect. During the product development process in particular, incorrect decisions are extremely critical because - depending on the industry sector involved - 70-85% of final product lifecycle costs are determined during this early stage.

All things considered, the status quo is characterized by the fact that the economic success of projects / products and therefore the success of many companies is mainly left to chance.

Some companies have already recognized the importance of integrated IT-systems in order to obtain a (cost) control over their products while passing through the lifecycle. They have built up complex IT-systems in recent years. However, for the most part, these IT infrastructures contain only a few or rudimentary PLM functionalities. Often these systems are nothing else than self-programmed extensions of ERP, CAx, or calculation tools, which have grown over the years.

Before cost-saving potentials are addressed in more detail, some basic preconditions for an efficient LCC Controlling are described, without which an appropriate identification of cost-saving potentials is not feasible.

Preconditions for an Efficient Product Lifecycle Costing

As already mentioned above, in order to explain the characteristics of an IT-supported product lifecycle costing, an ETO company product serves well as an example. Whenever there is talk of these kinds of products, it means that complex and unique products are involved. Here, the product development process entails long-term projects with many participants, requiring adequate co-ordination activities. Last but not least risk increases with project size, duration and product complexity. Usually such projects are built up within the ERP system as R&D or customer projects, serving as cost objects for the construction and manufacturing of the ETO (involving services and physical materials). In order to plan and assign costs and to execute scheduling activities for the project as a whole, including related components and services, it is essential to build up a work breakdown structure (WBS), which takes both lifecycle stages and product structures into account. An example-WBS can be found in Fig.2. Detailed structures also enable detailed reporting depth, but are very product-specific. Therefore standard structures are recommended with an appropriate level of detail. The advantage of these standard structures is that they allow for product portfolio comparisons, delivering important measures for the success of several similar products.

For example, the costs to be considered for the procurement of a pump (e.g. component 1 of a plant in Fig.2) not only include the acquisition costs, but also costs (mostly working hours, travel expenses, etc.) arising from sales activities, specification and design, delivery, transport … to name just a few!

After having built up the WBS (or better: copied a good standard one), cost planning activities follow. It should be possible to accommodate different revisions of cost calculations at a later stage to allow for planning and reporting of change costs versus actual change costs. Cost planning is usually based on a combination of prevailing cost information from the market and prior experience originating from similar past projects. These costs can be regarded as target costs.

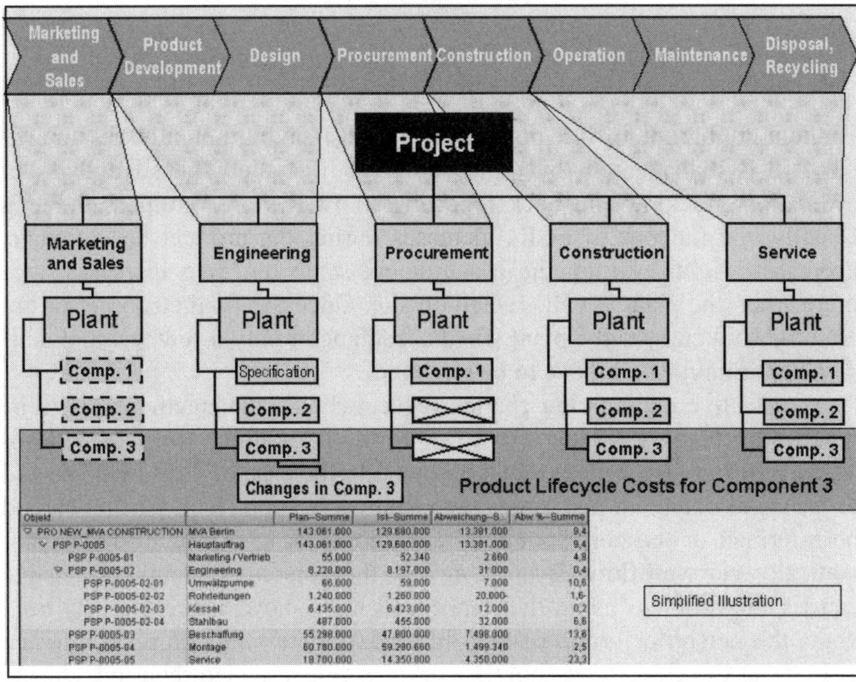

Fig.2: Gathering Product Lifecycle Costs of an ETO product in a structured manner using a Work Breakdown Structure (WBS) within the ERP system

The more that is known about the product characteristics and the market situation, the more detailed and exact are the cost forecasts / target costs and pre-determined budgets for the product/ components and the project itself.

Cost-Saving Potentials with an Efficient Product Lifecycle Costing

Experience gained from re-engineering of past projects enables an identification of cost drivers as well as the time of their occurrence. Thus, having detected the critical stages of the process, founded decisions can be made in time in order to take corrective action. Risk can therefore be reduced and cost-saving potentials achieved, as a result.

The subsequently documented cost-saving potentials refer, in particular, to collaboration aspects. They can only be recognized and estimated if they are based on an efficient product lifecycle costing supported by capable IT systems.

a) Engineering Change Request (ECR) Management During the Product Development Process (Design, Engineering, Procurement, Construction)

Without doubt, one of the most important cost-saving potentials is an efficient ECR management over the complete lifecycle. The advantages of evaluation, tracking and back reporting of ECR's are outlined in Fig.3. Usually, the number of ECRs increases during the product design stage, decreases slightly during the procurement stage but then increases once more when the plant is constructed on site. Once the final customer or operator has taken over the plant (final acceptance), only a few residual deficiencies remain which have to be removed.

An effective engineering change request (ECR) management has to be based on processes, which also allow for collaboration, not only with internal staff but also with external partners. In the case of ECRs with regard to technical details in particular, the responsible supplier or partner should be informed or the corresponding task should be assigned to them automatically via workflow. By this means, the person or group responsible can fix the problem using their specialist know-how. Once this has been done, the person or group responsible reports back the completion of the task, using the CPC/PDM-tool. If working hours are also reported, change costs can be calculated and compared with the planned change costs by WBS element. When using a WBS structure in the ERP system, appropriate standard ERP reports or a Business Warehouse (BW) system allow cost monitoring and controlling by WBS elements / components, including assigned change costs. Furthermore, a central change controlling team can monitor and trace the status of change requests and tasks within the project by responsible person or group and by LC stage. It is now possible to identify critical, change-intensive LC stages or components as well as to track down participants who frequently cause changes. In order to shorten change cycles, incentives can be given to participants who complete their tasks on time or by reallocation of resources. Shortening change cycles save costs and avoids penalties, caused by not keeping to within set deadlines.

Product changes during construction or warranty are usually extremely expensive. That is why it is important to avoid changes at all or - if they cannot be avoided completely - to move them to the beginning of the life cycle and fix them as soon as possible. This enables sequence errors and consequential costs to be kept to a minimum. As a consequence of a good ECR management, product quality can be improved. Ultimately, this leads to an extension of product life and potentially lower maintenance costs.

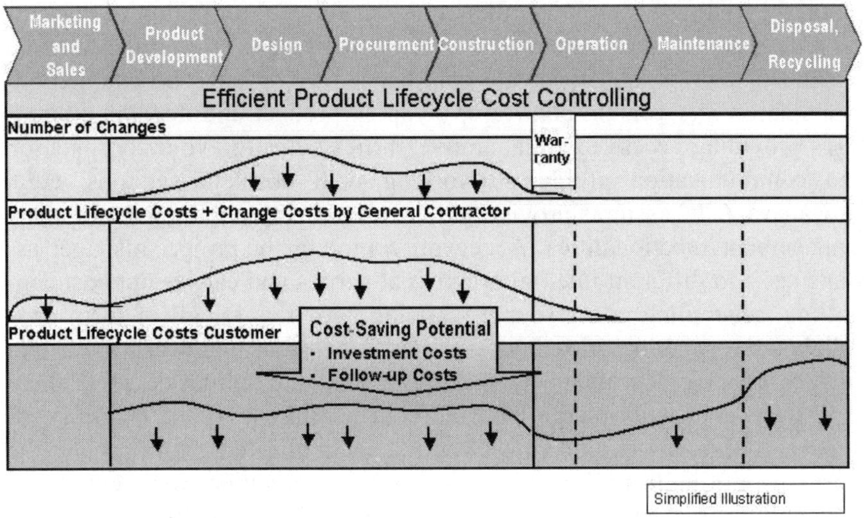

Fig.3: Identification of Cost-Saving Potentials regarding ECR management (e.g. by collaboration) with the help of Product Lifecycle Costing

b) Design-to-cost – Consideration of Cost Aspects in Product Design Stages

Typically, engineers develop complex products, taking only technical aspects into account. Some design-to-cost aspects, which can be easily applied, are demonstrated below:

For example, a purchasing department may, on the one hand, add target prices to the material master data. Purchasing departments usually get this kind of information from basic agreements with suppliers for catalogue parts/components or standard services or from other agreements. Even if the purchasing department negotiates lower prices, these numbers are sufficient for cost estimation during product design. If this basic cost information is provided, engineers can compare different product configurations, not only by functionality, but also by cost. They can therefore select the cheaper solution of two product designs if they do not differ substantially by functionality from each other.

On the other hand, design-to-cost can take place when the engineering department reuses standard components whenever possible. Databases with standard parts and components should be used in CAx or PDM-tools. Here, easy access with comfortable search engines is essential.

Last but not least, when talking about design-to-cost, the collaboration aspect also plays an important role. CPC systems enable efficient utilisation of suppliers' and development partners` product know-how. Furthermore, they allow for an integration and transfer of information of customers with regard to the entire product development process, in order to avoid misunderstandings and to create confidence. A CPC system-enabled administration and version control of complex product and document structures providing access to collaborative parties helps to avoid coordination and communication problems (working with obsolete versions, etc.). Nowadays, CPC systems provide full collaborative programme and project management functionalities. A general contractor or project manager assigns tasks to different internal or external parties and carries out cost controlling, scheduling and resource planning activities for all of them on a collaborative platform. Methods such as concurrent engineering (several engineers tackle a problem by splitting a task into sub-tasks, solve them separately and finally merge their individual solutions to one common solution), continuous engineering (working without interruption around the world, benefiting from time zones and modern communication technology) and simultaneous engineering (completing tasks in parallel) can be applied easily and for all participants.

All in all, optimisation in scheduling helps to shorten development cycles and again saves costs.

c) Operation and After-Sales Service

For the operator or final customer, the LC stages between the commissioning and plant disposal are especially important because during these stages the plant has to generate its benefits in the form of revenues. The initial investment analysis, which has been used to justify the construction of the plant, will be the benchmark for comparison with the actual performance of the plant. High product quality and professional after-sales service increase the number of operation years and with it the revenue generated in operation. Well-directed, structured and standardised communication of field experience data into the relevant preceding stages of the LC can avoid errors in the new product development process and furthermore reduce plant downtime. The functionalities are provided by maintenance management systems (MMS). Service personnel can fill in the field experience forms via the Internet from the site. This data can be transferred to the collaboration software (CPC-system) which itself has interfaces with the ERP, CRM, SRM systems. Thus all involved parties have prompt access to the relevant field data and can initiate corrective action (block master data to avoid reuse, improve product functionality, etc.).

d) Reporting

Powerful lifecycle costing depends to a large extent on a transparent and agreed reporting system. In order to obtain a detailed overview over the project / product structure, standard reports should include the following information:

- The actual state of the product and related objects at any time within the LC (actual status, overview of current billing and purchase orders)
- Status, progress reports and number of open and completed tasks and ECRs by duration of processing, responsible person/group, project / product structure / period
- Cost information: actual versus planned costs and budgets, change costs per project / product structure / period
- Product portfolio comparisons

An efficient PLM system should also be able to generate automated workflows, following predefined processes in order to inform initiators, responsible persons or other relevant groups whenever deviations from the normal product lifecycle process occur, e.g. actual costs exceed planned costs or budgets, deviation from pre defined schedules, ECRs not completed within a certain time span, etc. These kinds of signals, which are generated automatically by the system (warnings, information notifications, error notifications), give those responsible the possibility to take corrective action as soon as possible. Furthermore, determined and well-researched decisions can be made at the right moment. Of course, a prerequisite is a detailed customisation of workflows within the CPC-/PLM-tools and/or within integrated systems.

From an organisational viewpoint, LCC controlling should be executed on a continuous basis by product or project controllers or managers, even if the workflow generates automated reports. As a connecting link to the top management within a company, a central department should be responsible for product portfolio reporting in order to measure the success of several similar products. Results should be coordinated with the overall product strategy and corporate strategy.

The cost-saving potentials mentioned above can be seen as the most considerable ones. However, there are definitely more of them or related ones. All in all, these cost-saving potentials can only be realised if all participants involved are provided with a timely, easy and high quality access to consistent project and product data, in keeping with their role and authority.

IT Solutions

Modern PLM and CPC systems allow a role-based and workflow-orientated definition process control. The participant roles can be spread all over the world within and beyond company boundaries. These processes cover PLM functionalities such as collaborative project management, engineering change management, product configuration, document management, version control of products and documents using collaborative platforms accessed via the Internet. That is why they are also called CPC tools. PLM-/CPC tools not only facilitate collaboration with external partners, but also provide wide integration with CAx-, CRM-, EDM/PDM-, SRM-, ERP- and BW-systems - taking data security aspects into consideration, of course. Access control to product-related data can be structured depending on the detail of the product or project structures chosen. It is possible, therefore, to protect certain data in order to avoid abuse. BW systems are normally used for internal company purposes. BW systems can generate reports from various data, which are processed in any connected systems. Portal solutions represent the platform beyond the company borders, giving external partners the possibility to communicate with the company. Sensitive data is protected by applying structured access control and firewalls.

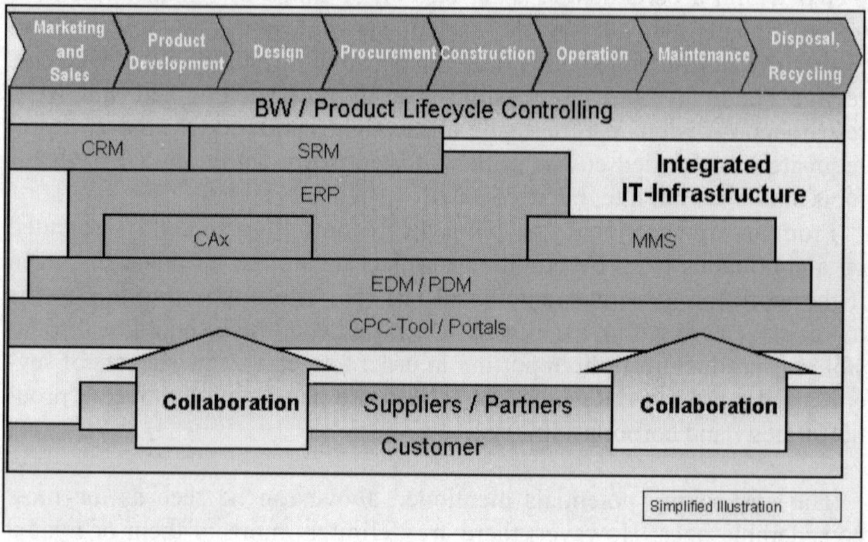

Fig.4: Integrated IT systems – Precondition for Efficient Product Lifecycle Management

Fig.4 gives an overview of the most important IT systems, including the interrelating interfaces. It also shows the relevance of different IT systems by product lifecycle stage.

Conclusion

Past product and project experience and actual market forces determine the basis for target cost estimations.

In order to assign actual and planned costs, change costs or budgets to products and projects, it is useful, especially for ETO companies, to build up project structures according to the product lifecycle stages and the product structure in the ERP system which is the basis of cost information.

An efficient product lifecycle costing allows for identification of cost drivers and cost-saving potentials.

One of the most important cost-saving potentials can be seen in the collaboration with external partners, suppliers and, on demand, customers. CPC solutions combined with the integration of traditional IT-systems can make them work together as one team. CPC tools are not only able to distribute information, but can also be used for assigning and tracking tasks.

The shortening of the product lifecycle and associated cost savings caused by collaboration are characterized by the fact that they do not happen at the expense of product quality. On the contrary – product quality is improved, because technical knowledge is provided by the corresponding specialists, who participate in the collaborative project. Continuous monitoring and analysis of the business and distribution of information generate extensive learning effects.

Well-designed processes, structured organizations with the appropriate responsibilities, combined with frictionless integrated IT- systems assist companies in solving complex technical problems without losing the cost focus. These are the best prerequisites for achieving a long-term competitive advantage.

Glossary

Abbreviation	Meaning
BW	Business Warehouse, central reporting system which uses information from various integrated IT-systems
CAx	Computer Aided … (x for design, engineering etc.)
CPC	Collaborative product commerce
CRM	Customer Relationship Management
ECR	Engineering change request
EDM / PDM	Electronic / product data management
ERP	Enterprise Resource Planning
ETO	Engineer-to-Order (Investment Goods, e.g. power plant engineering and construction
IT	Information Technology
LC	Product Lifecycle
LCC	(Product) Lifecycle Costs
MMS	Maintenance management system
PLM	Product Lifecycle Management
SRM	Supplier Relationship Management
WBS	Work breakdown structure (used for complex projects)

Competition and Collaboration in the Automotive Industry

Klaus Kaplaner

Nexolab GmbH

In a time when markets are becoming increasingly saturated, automotive manufacturers are faced not only with the necessity to provide a wide variety of vehicles to satisfy market demand but are also developing products which are increasingly complex from a technical viewpoint – taken together difficult tasks to deal with at the same time. Over and above this, legal, cultural and economic factors are constantly shifting ground and, due to customer demand, product innovation cycles are having to become ever shorter. In short, the challenges facing car manufacturers have never been greater than they are today.

Collaboration is seen as the new form of cooperation between individual companies in order to deal with such challenges. Collaborate or compete is the newest term for earning revenue.

The question needs to be asked whether or not cooperation is already being practiced in today's economy. Aren't cars already being assembled and isn't their make-up to a greater or lesser extent being determined by automotive suppliers and their shares of the total production value of a vehicle? What will really change in the future? In short, car companies are faced with a barrage of customer's demands for even more varied, personalised vehicles and for these to be delivered as soon as possible.

In Germany alone, in comparison to 10 years ago, the number of product variants has doubled, product development times have been cut in half and the time from order to delivery has been reduced to "but a few days or weeks".

New electronics have increased the technical complexity of vehicles and increasingly customer's product and service desires are being addressed through expanded offerings.

Different disciplines are growing together as the product & service offering of automotive companies are expanding to meet consumer demands. The automotive company of the future may well offer (or need to offer) product development, production, financial services, travel, transportation & mobility, entertainment & communication services and all-encompassing customer management facilities; to name but a few!

But is this likely to be the case? Even today, we can see that such large, inflexible companies who try themselves to cover the whole spectrum of customer needs are especially vulnerable to changes in market conditions and can become unstuck by a simple change in the competitive environment.

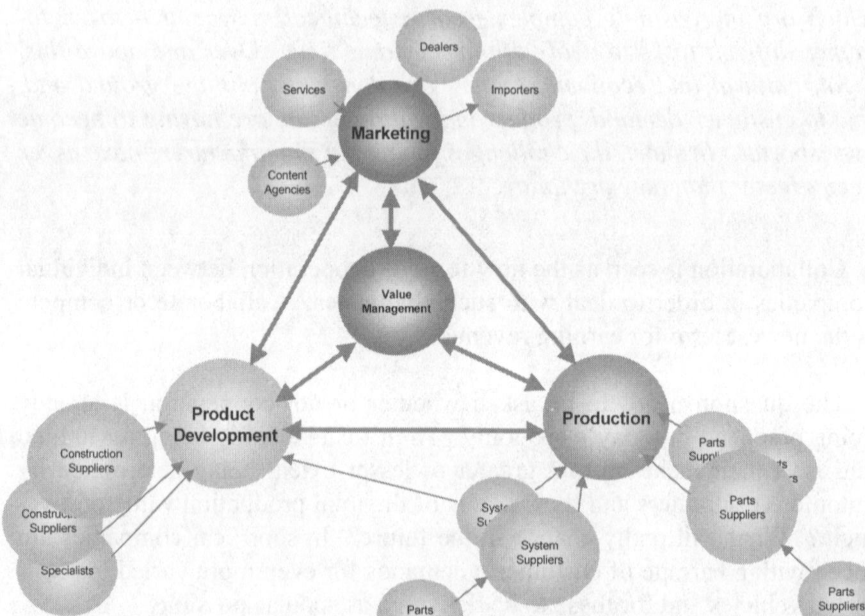

Fig.1: Future Industrial Model: Cross Company Collaborative Networks

The classical company which has its focus on development, production, sales & marketing will soon be a relic of the past, in our view. Even today, it's very clear that companies who manage to achieve a clear focus on their core competences can race beyond being considered one of the many average industry performers. This not only applies to the automotive industry

– look at Adidas whose competence lies in marketing & branding or even Dell Computers where it is marketing and integrated logistical operations.

Such cooperative arrangements are already well-known in the automotive industry

- BMW is developing its new X3 sports activity vehicle together with the system supplier, Magna Steyr and the vehicle is to be manufactured in a Magna-owned factory
- The final assembly of Porsche's Boxster is done by Valmet in Finland
- Volkswagen's Tuareg and Porsche's Cayenne share a common platform, and
- PSA and BMW have started a joint project to develop and produce a new family of engines for future variants of the Mini.

The current value of work on the car done by suppliers lies at around 60 to 70% of the production cost and this percentage will increase. Collaborative networks of companies focusing and specializing in their core competences are here now and are here to stay – developing, producing and marketing products jointly and efficiently.

But these networks may not last forever. It doesn't mean that what works for one particular product or product line will necessarily work for another. This means, of course, that, in one particular case, companies may cooperate, in another they could be competitors. Compete & collaborate will become the new watchwords on the way to making money!

Whilst focussing on strengths and core competences may allow companies to offer customers the highest levels of service and delivery of products within shorter timescales according to their demands, there are still challenges to be dealt with. The advantages of companies collaborating may well be negated as a result of inefficient working processes, badly-functioning information sharing and an ever-increasing need for coordination work. Differing ways of working, different IT systems, product groupings and use of reference numbering can lead to chaos. Effective collaboration means having an agreed infrastructure made up of processes, systems and rules of the road. Marketplaces and platform developers are candidating to run operational processes efficiently and to standardize these wherever possible.

Using external consultants and training companies may also help but they will be required to be much more subject matter experts than was the case until now. In short, industry expert and specialist advisor in one! That integrity and intercultural abilities are pre-requisites goes without saying.

A further issue is the economics of collaboration. Collaboration means partners jointly sharing both advantages and risk. Suppliers may in future see themselves involved in sharing the marketing risk for new products (will they be accepted by customers?). Risk and benefit sharing models need to be clearly defined before new collaborative efforts are undertaken and the consequences of these need to be clearly understood by all parties.

Using third parties would also seem to be a good approach here – calculating business cases, proving these and providing a neutral partner for the organizations involved is critical in my view.

If the aspects I have mentioned above are taken into account, then there is no reason why companies cannot deal effectively with both today's and tomorrow's challenges. It is clear to all of us, however, that such networks are in no way fixed and immovable and they may only exist for as long as a particular project lives on. It is both feasible and possible that today's competitors will be tomorrow's companies who will collaborate with one-another. The choice between competing or collaborating will not be that which has a considerable influence on the future business landscape and daily events, but rather the issue of competing & collaborating. Indeed, accepting this and implementing appropriate strategies is the real challenge facing companies today.

Best-Practice Concept for Supply Chain Monitoring

Oliver Merle

Odette International Ltd.

Modern Supply Chain Management (SCM) concepts address important root-causes of current limitations in logistics networks. Efficient built-to-order strategies and short time to delivery goals require fast and flexible supply networks. In this context inter-company visibility and transparency *of relevant data play a central role. The promising potential of modern SCM concepts is attracting the attention of the whole automotive industry.*

Modern SCM is targeting to overcome *the* local, company-centric view *by* looking from a global perspective on value adding networks.

Supply Chain Monitoring *(SCMo) is a multi-level SCM concept supporting the fulfillment process. It helps to better master critical supply networks by checking permanently if the actual inventory in the observed supply network is compliant (synchronized) with the respective demand.*

The first SCMo concepts have been implemented more than 2 years ago. Today more than 100 companies (carmakers and suppliers of modules, systems, components and parts) are using SCMo in the day-to-day business and every months new supply networks are added. The members of the Odette SCM group[1]*, i.e. key-players on OEM and supplier level, brought their know-how and experience together and documented it.*

This paper summarizes the Odette Recommendation for Supply Chain Monitoring (published in April 2003), which shall become the best practice standard for future SCMo concepts and the corresponding IT-applications.

---

[1] Audi, BMW, DaimlerChrysler, Ford, GM, PSA; Renault, Volvo, Volkswagen, Bosch, Faurecia, SiemensVDO, Treves and National Organizations of the European Automotive Industry

1 Philosophy of Supply Chain Management

Starting point of Supply Chain Management activities is an in depth understanding of the current situation in the supply network (e.g. material and information flows). Based on this, consistent and agreed cross-company business processes are established. In general this requires an adaptation / reengineering of internal business processes of the participants. IT and the Internet play a major role to enable and support SCM concepts.

Core elements of the SCM philosophy are:

- Customer demand drives the whole inter-company supply network (*synchronization, built-to-order*)
- Increased reaction speed and *flexibility* of the supply network
- Where applicable *multi-tier concepts* are enabled
- *Exception* based management (alerting) based on workflows
- Integrated inter-company processes are needed to collect and share relevant data
  → Visibility and transparency
- Information relevant for decision making is updated in an appropriate period of time
  → quick responses
- *Scenarios* can be simulated due to the global availability of information
- Enable *win-win* partnership
- Find *root causes* rather than curing the symptoms

2 Introduction to Supply Chain Monitoring (SCMo)

SCMo applications are an add-on to existing ERP-systems, which remain the leading systems for every organization. SCMo provides a functionality that cannot be achieved by decentralized ERP-systems. A key success factor for SCMo is to keep the complexity low (e.g. number of parameters, detail of modeling) as the main benefits come from speed of information flow and cross-enterprise visibility and synchronization. The general principles are:

- *Integration* of supply chain participants in the short / medium-term fulfillment/execution process by exchange of demand and inventory information plus eventually status of production orders (future inventory)

- *Transparency* and *Visibility* of inventories and demands for multi level supply chains due to close to real-time updates from decentralized back-end systems
- Eliminate *time lags* in the information flow
- *Synchronization* of demand information along the *critical* paths of selected supply network (see Fig.1) by *fast demand calculation* throughout the selected parts of the supply network
- *Monitoring* of the selected supply network part (early warning functionality and exception based management)
- *Instant feedback* on impact of decision for every participant of the supply network
- Smart, fast, transparent concept with limited complexity

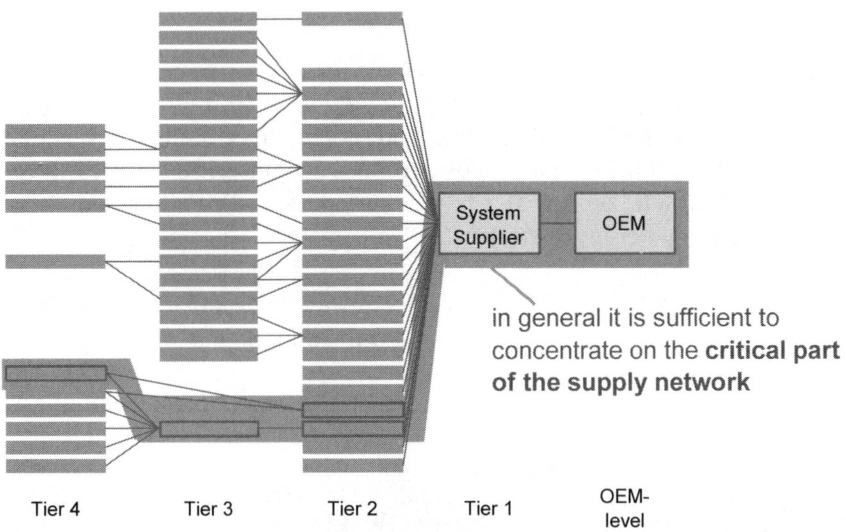

Fig.1: Example for critical part of supply network

In the following the term "*network*" is used for the critical part of the total supply network.

Goals of SCMo are:

- *Prevention of bottleneck situations in critical supply chains* („feasibility of short term forecasts / releases") by early warning functionality (alerts)

- Network-wide optimization due to network-wide inventory transparency and elimination of bull-whip-effect
- Optimised *allocation* in bottleneck situations due to network-wide inventory and demand transparency enable better allocation with less administrative effort

Benefits of SCMo are:

- Improved inventory levels in the selected part of the supply network.
- Increased flexibility and synchronization of the supply network.
- Reduction of non-value adding cost (trouble shooting, administrative effort to manage and control material flow, etc.).
- Reduction of premium freight.
- Avoidance of scrap due to obsolescence (engineering changes, end of production).

The basic steps to set up and initialize SCMo are:

- Determination of the potentially *critical part of the supply network* Criteria are high shortage risk and effect, long lead and reaction times, high total inventory cost, frequent engineering changes, creation of variants on low tier level, bottleneck and/or allocation situation at a low tier level, etc.)

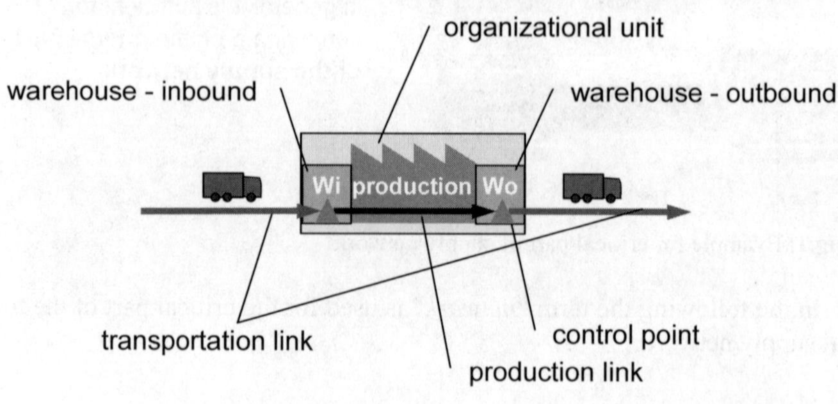

Fig.2: Network-elements of SCMo model

- Mapping of the *structure* of the selected network to the *SCMo model* (mathematical graph).
  Control points represent warehouses (inbound, outbound, consign-

ment stock, internal warehouses, etc.) and represent the nodes of the graph. Transportation and production links represent edges.
- The result is a computerized model in the SCMo application. (see Fig.3 below)

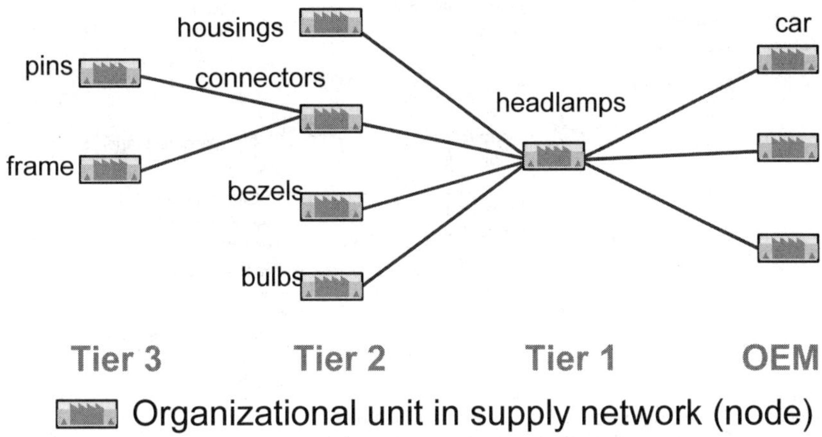

Fig.3: Example for visualization of critical part of supply network

- Entry of parameters for the SCMo model:
  → Organizational units (e.g. contact persons, organization ID)
  → Control points (e.g. minimum/maximum inventory levels)
  → Production link (e.g. production lead times, bill of material)
  → Transportation link (e.g. transportation lead time, part number reference table)

After completion of these steps, the SCMo system is ready for use. The basic activities and functionality in the day-to-day business are:

- Regular update of *inventory* and *demand* information,
- *Alert* generation when a monitored figure is going out of the predefined range (=exception)
- *Transparency* allowing the user to easily understand the exceptional situation due to suitable presentation of the relevant data (alert board)
- Counter measures are documented (action item board) and monitored by a *workflow* concept

Today's SCMo systems use a centralized architecture. That means, a central application is hosted on a central server.

3 Business Process Description

The figure below shows the basic idea of the SCMo concept.

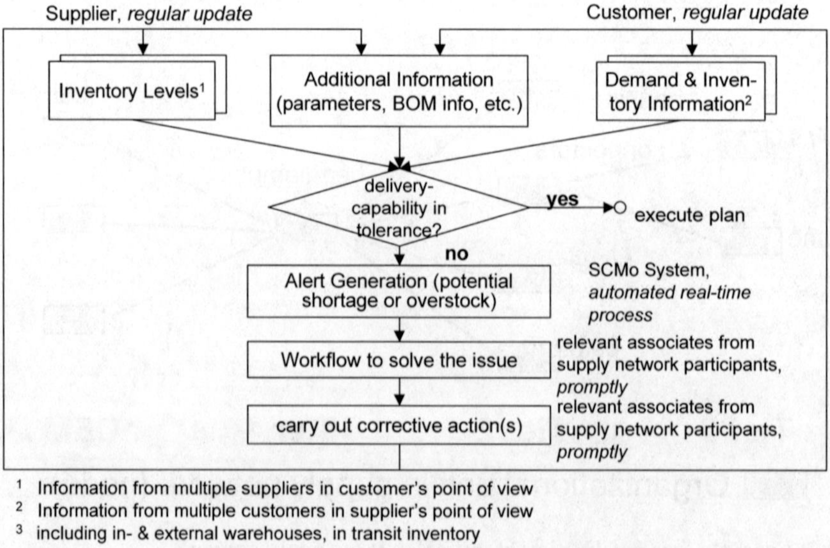

Fig.4: Basic idea of SCMo process

The Odette SCMo Recommendation includes a detailed business process description that exceed by far the scope of this paper.

- The *Initialization Processes* describe how to identify the relevant part of the supply network, how to model it in the SCMo application and to set up the complete concept.
- The *Operative Processes* describe the demand update, inventory update, alert calculation and alert management. The basic idea, presented in Fig.4, is detailed in this section.
- The *Service Processes* describe change of master data and parameters, change of alert parameters, measurement tracking and reporting.

4 Functionality

4.1 SCMo-Model

The computerized SCMo model is build up with only 4 basic types of elements as shown in Fig.2:

An *Organizational Unit* is representing in general a plant. In most cases an organizational unit has 2 control points (inbound and outbound warehouse); however it can also have only one or more than 2 control points

A *Control Point* represent a physical warehouse, e.g. inbound and outbound warehouse, internal and consignment stock warehouse, etc. The alert calculation is based on control point information, e.g. inventory levels and demand. Control points are nodes in the supply network graph.

A *Production Link* connects 2 control points within one organizational unit. It describes the relationship between 2 control points on part number level, representing a value adding process (e.g. control point to control point BOM, production lead time). It is every company's internal responsibility to create and regularly maintain the parameters.

A *Transportation Link* connects 2 inventory control points of different organizational units. It describes the relationship between 2 control points on part number level, representing a transportation process (e.g. mapping of supplier and customer part numbers / cross-references, transportation lead time). The responsibility for creation and regular maintenance needs to be clearly defined between the 2 partners.

4.2 Supply Chain Modelling / Mapping

We recommend to install a central contact person for every supply network. It should be in charge of analyzing and identifying the critical part of the supply network together with contact persons of other organizational units. After the participants have agreed to establish the SCMo concept (we recommend kick-off meetings with logistics and IT), an efficient and safe way is needed to computerize the SCMo model. That means the complete SCMo model with all its parameters needs to be built up. The proposed decentralized, iterative approach guarantees that every organization keeps full control on its business. It also gives full responsibility to every single organization for correct and updated data to ensure a sufficient data quality. The central contact person will support this process.

Iteration 1:

The initiator starts the process with "his" part of the SCMo model (step 1 and 2 in the Fig.5) and involves its direct customers and/or suppliers by sending them e-mail with a proposal for the transportation link information. With step 3 the direct customers and suppliers check and complete this proposal.

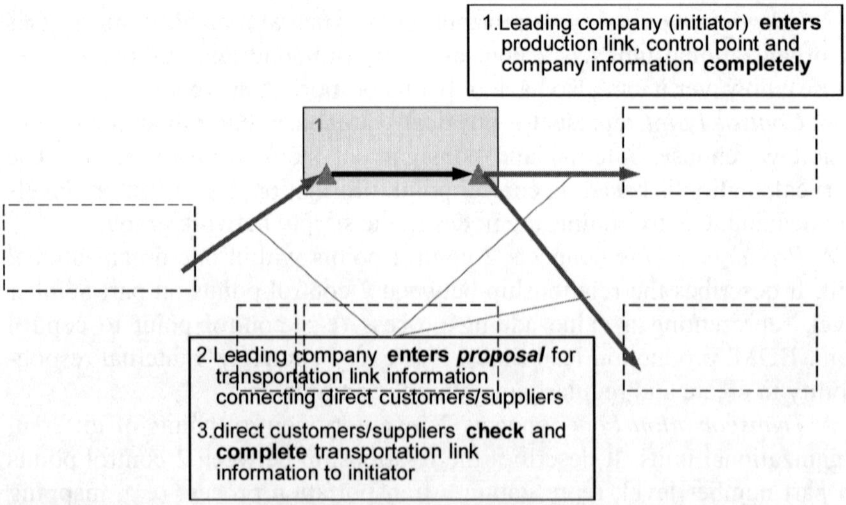

Fig.5: Supply chain mapping / modelling – iteration 1

Iteration 2

The direct customers and/or suppliers of Iteration 1 take the role of a leading company. The iterative process starts again.

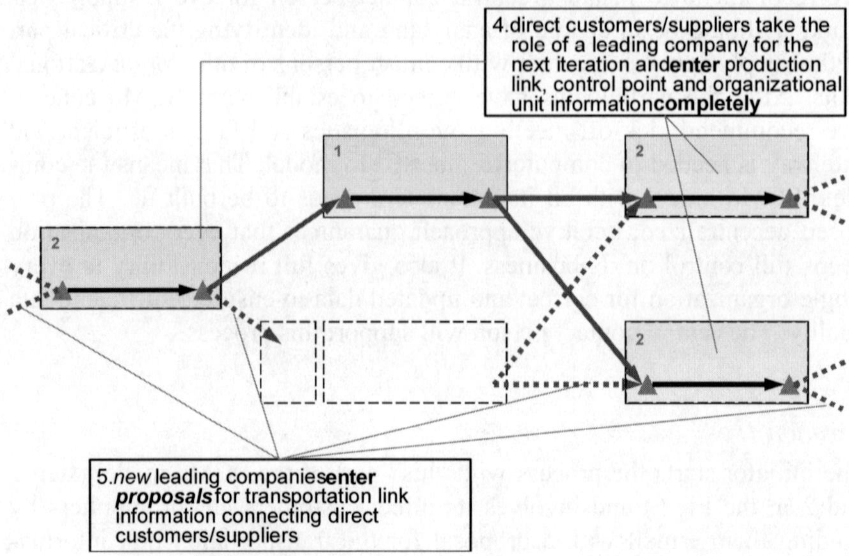

Fig.6: Supply chain mapping / modelling – iteration 2

The iterative process stops when the complete structure of the selected network is modeled.

The detailed process description of the initialization phase can be found in the html-package.

4.3 Inventory Definition

In general, company internal backend systems classify inventories, e.g. in disposable, allocated / reserved for customers, blocked, in quality inspection, rejected, to be reworked. In SCMo only *available* inventory is used. The available inventory can be either already allocated / reserved for a specific customers or for free disposition.

A definition has to be made company internally what classes of inventories will be considered as *available* inventory in the SCMo application.

Fig.7 shows the different types of inventory considered in SCMo.

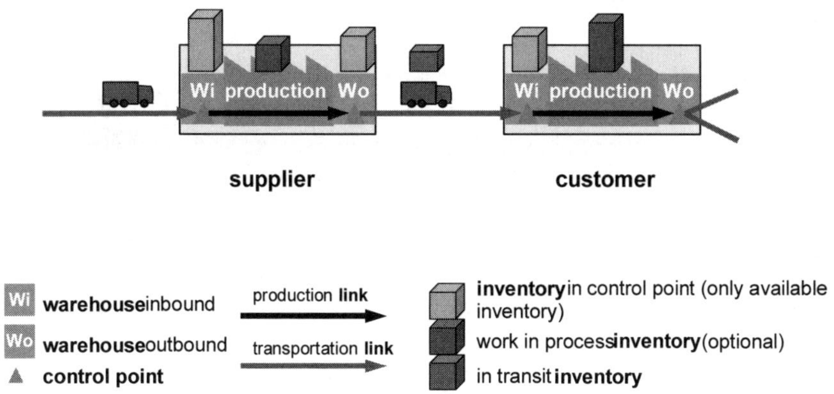

Fig.7: Inventory definition

For alert calculation (and for demand calculation) the in-transit and work in process inventories need to be allocated completely to the respective ship-to and ship-from control points. There are several strategies possible. The strategy for in-transit inventory allocation must be agreed between supplier and customer.

In the *standard allocation strategy* the in-transit inventory will be allocated to the ship-from control point (outbound warehouse) and the work in process inventory will be allocated to the ship-from control point (inbound or internal warehouse). This strategy equals the well-known back flushing method for work in process inventory.

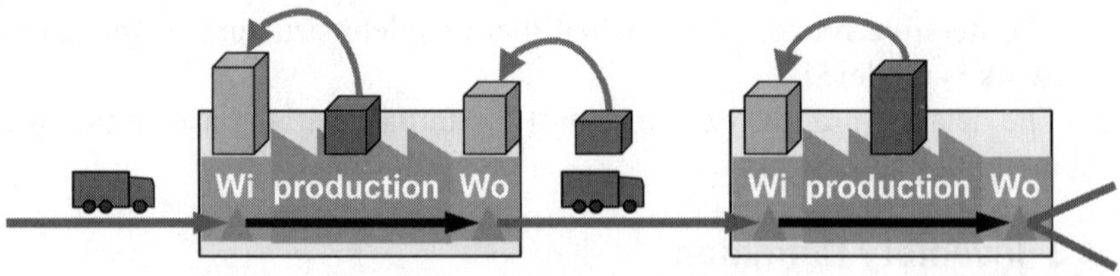

Fig.8: Example for standard inventory allocation strategy

In transit inventory should be calculated by matching delivery notes / ASN with goods received, as shown in Fig.9.

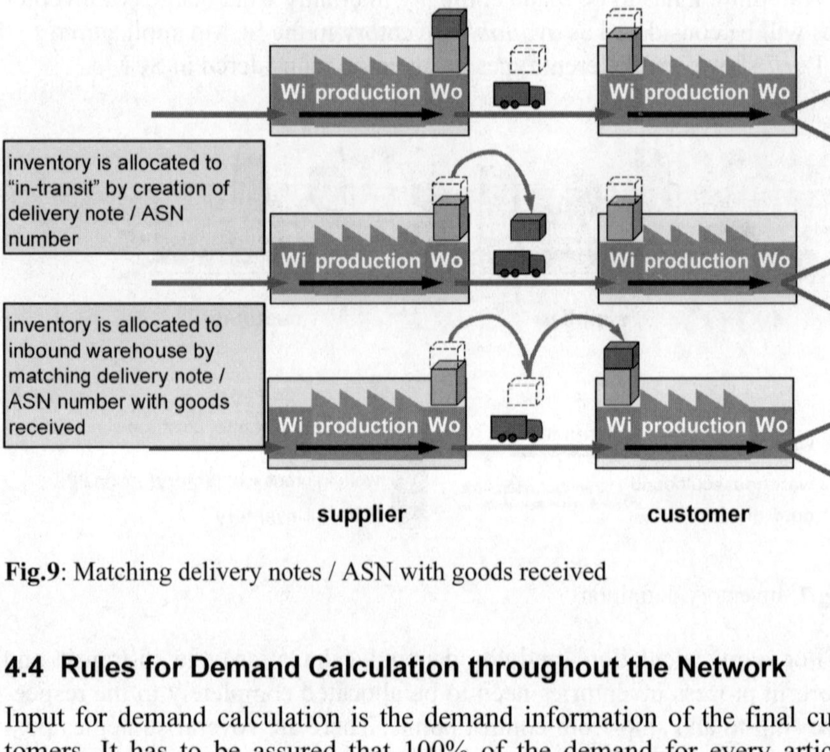

Fig.9: Matching delivery notes / ASN with goods received

4.4 Rules for Demand Calculation throughout the Network

Input for demand calculation is the demand information of the final customers. It has to be assured that 100% of the demand for every article number is reflected in the SCMo system. I.e. additional demand from companies that are not modeled in the SCMo system (e.g. spare parts, CKD) must be included. Manual input of additional demand must be possible at any tier. A production calendar and a transportation calendar are needed for demand time planning.

As shown in Fig.10, the several demand calculation steps in today's in-house-systems of every organizational unit (in general ERP-systems) can be assigned to the network elements of the SCMo Model. The calculation goes

- From the outbound control point
- To production link, e.g. using the bill of material
- To the inbound control point
- And finally connecting to the next organizational unit using a transportation link

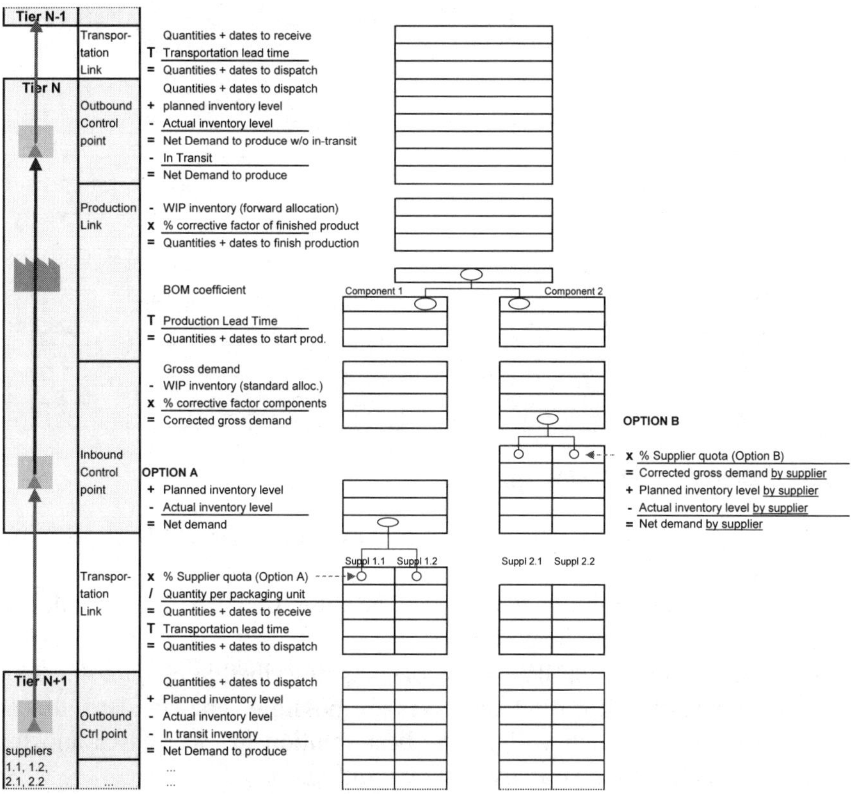

Fig.10: Basic demand calculation scheme

Demand is propagated recursively over to the next tiers of the supply chain as illustrated in Fig.11.

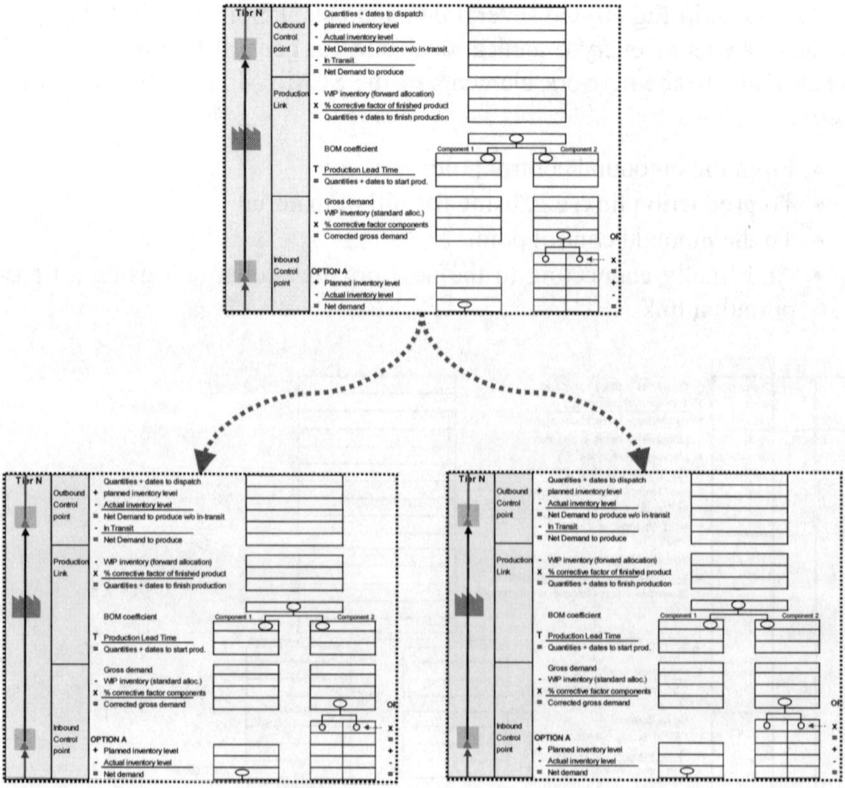

Fig.11: Recursive demand propagation

The basic calculation scheme shown above does also apply for SCMo-networks.

There are several options that need to be setup properly in the modeling and reconciled between the partners:

- *Consideration of the WIP inventory:* As described in chapter 4.3, several inventory allocation strategies are possible. The standard allocation strategy, which will be the best solution in most cases and the forward allocation strategy is shown in Fig.10.
- *Consideration of supplier quotas* (only relevant for multi-sourcing scenario):
  → Option A: In the MRP-run, the net demand for components is calculated by part number, not considering the demand for individual suppliers. I.e. supplier quotas are applied after the net demand calculation (see example with component 1)
  → Option B: In the MRP-run, the net demand for components is calculated by supplier and part number. In addition the inventory

information must be available per supplier. In this case, the quotas can be applied after the GROSS demand calculation (see example with component 2).
Option B enables a VMI concept with several suppliers for the same part number.

- **Definition of the handing over point of the demand information / scope of responsibility of the partners**

  Fig.12 shows 3 examples for different delivery terms and/or fulfillment methods that lead to different types of demand information. A demand type identifier is needed to communicate the exact meaning of the demand figures between customer and supplier:

  - The **EXW** incoterm scenario is shown for supplier 1.1:
    The customer sends dispatch dates and quantities, i.e. net demands with forward termination.
  - The **CFA** incoterm scenario is illustrated for supplier 1.2:
    The customer sends receiving dates and quantities, i.e. net demands without forward termination
  - The **VMI** scenario is shown for supplier 2.1:
    The customer sends corrected gross demands.

Depending on the contract the area of responsibility of each partner is defined, as shown in Fig.12 (other cases are possible)

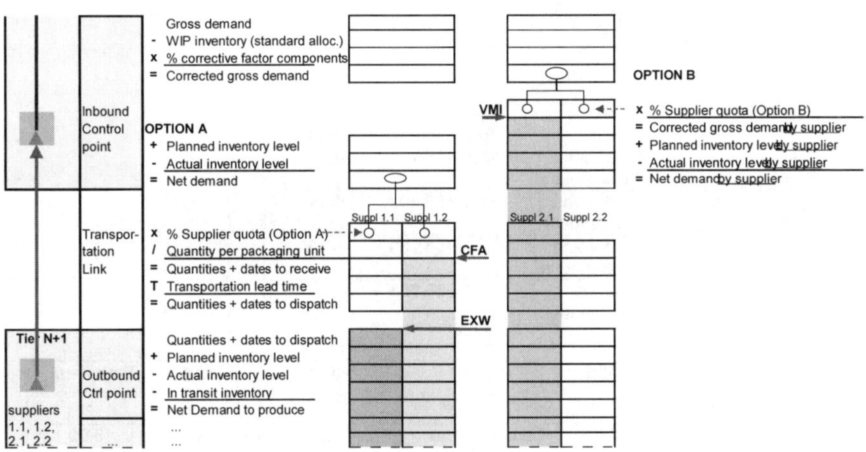

Fig.12: Area of responsibilities of each partner

4.5 Alert Management, Communication and Workflows

SCMo supports the principle of exception-based management. That implies, that the system needs to recognize if a situation (e.g. demand / inventory) is "normal" or "exceptional". Therefore thresholds need to be defined (e.g. minimum and maximum days of inventory).

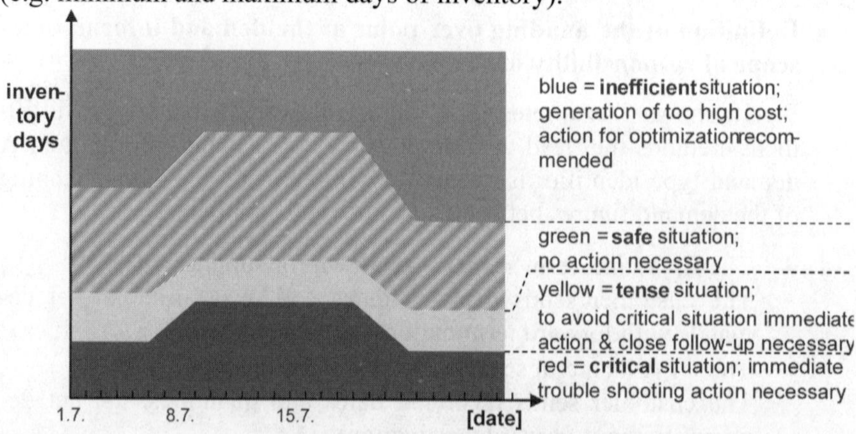

Fig.13: Alert color definition for "days of inventory alert"

Alerts inform the respective persons in charge about those exceptional situations, which are visualized in the alert board / action item board of the SCMo application (see Fig.14 below)

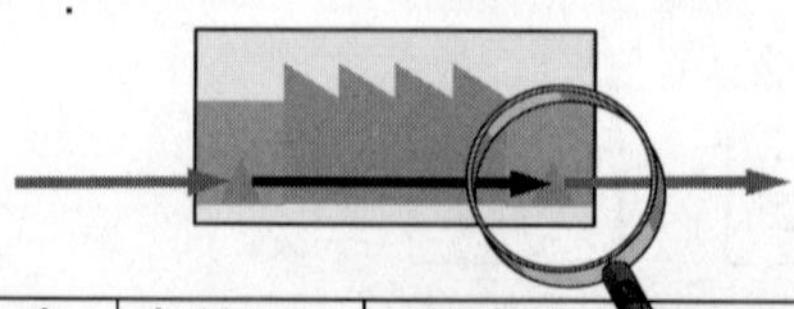

color	alert type	control point	partnumber	action item
● (red)	low inventory	outbound warehouse	123 456 701	A 17
● (red)	low inventory	outbound warehouse	123 456 702	A 17
● (yell)	low inventory	outbound warehouse	123 456 703	A 13
● (blue)	high inventory	outbound warehouse	123 456 706	

action item	step	Who	Action	When	Status	Follow-up
A 17	1	John	Call maintenance	10:00	Done	-
	2	Maint	Inform Repair	16:00	In progres	15:00

Fig.14: Alert generation, alert board and action item board

Filtering, sorting and grouping of alerts support the end user. Criteria for filtering and sorting could be priority, alert type, alert color or user. *Alert notifications* (e.g. emails) are generated and distributed to predefined user group(s) including escalation.

The following *standard alerts* have been identified:

- Low (excess) inventory / days of inventory
- Significant change of demand (e.g. new vs. previous information or standard deviation)
- No data update for more than x days/hours
- Engineering change notification
- Late/no arrival (e.g. no matching of delivery notes / ASN with goods received within standard time frame)

Communication of alerts is restricted to the *relevant* part of the supply network.

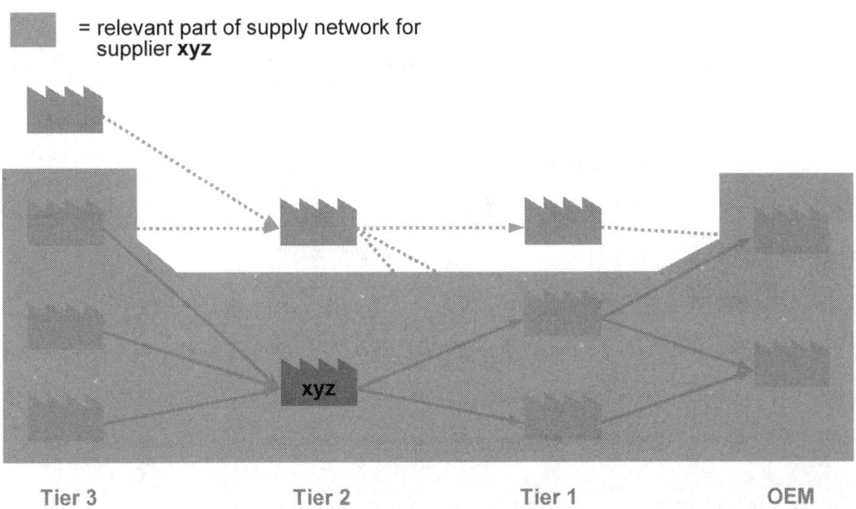

Fig.15: Example for relevant part of supply network

Workflow Functionality supports and controls the problem solving process. The measures from the action item board are controlled according to the defined review / completion dates and times and persons in charge.

4.6 Multi-Customer Scenarios

Multi customers scenario means that a supplier delivers the same part to several customers. This scenario can appear multiple times at every tier level.

The challenge is to give correct alert and inventory information to the individual customers.

In the approach without allocation of inventories, the *total inventory* is compared with the *total (aggregated) demand.* Every customer sees the total inventory at the supplier and its total demand. The alert definitions cannot be customer specific, i.e. all customer are getting exactly the same alerts.

In the approach with allocation of inventories (preferred), the *customer allocated inventory* is compared with the *individual customer demand.* Every customer sees only the part of the inventory which is allocated to him. For every customer-supplier relationship alerts are defined individually.

Inventory allocations are based on input from backend system or allocation strategies implemented in SCMo application.

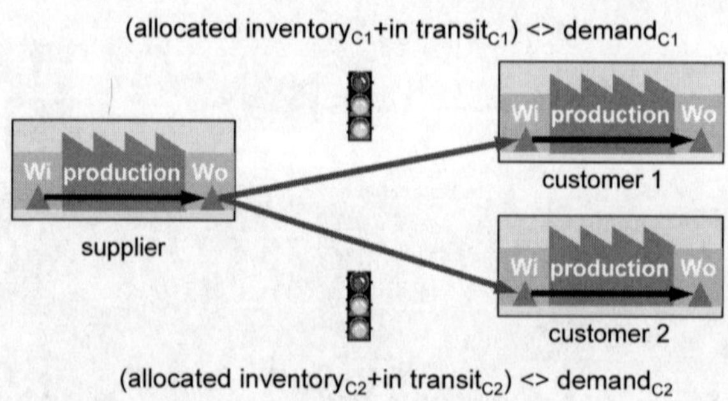

Fig.16: Multi-customer scenario with inventory allocation

Allocation Strategies

Only if no inventory allocation is available from backend system, an automatic "virtual" allocation in the SCMo system should be considered. The allocation strategy should be mutually defined by supplier and customer. Allocation strategies can be based on quotas (e.g. fixed quotas, demand

mand quotas, capacity quotas) or priorities (high priority for OEMs, low for after market).

The allocation strategy should try to satisfy the minimum requirements of all customers. Current inventory levels should be considered and minimum inventory levels should be reached before allocation strategy starts.

4.7 Creation and Maintenance of Parameters and Management of Engineering Changes

It is fundamental to update the SCMo model whenever there is a significant change in the real supply network. The SCMo application should allow *manual* and *automated* creation and maintenance of parameters. The leading system should always be the backend system.

In order to keep the SCMo model consistent, updated and close to reality, support functionality is required.

Management of Engineering Changes will still need manual interaction. However it is supported by clear *visualization* (table or graph) of part number references to enable easy checking and a *completeness check functionality* displaying every supplier part number without corresponding customer part number and vice versa (logic check). The system shall generate automatically alerts (displayed in alert boards plus e-mail) for all effected direct customers/suppliers. The persons in charge have to confirm actively that the effects of the engineering change are completely checked and all parameters are completely updated.

In order to avoid excess manual work a *replacing functionality* (mass change) is required. Example: all part numbers x in the SCMo application change to part number y.

4.8 Calculation and Statistics of KPIs

The following KPIs could be calculated by the SCMo application:

- Customer demand forecast quality
  (How "good" is the demand forecast I got x weeks prior to fix call off / release?)
- Inventory days on hand (by control point and aggregated over several control points)
- Number of parts (and time) out of safe/green range (alerts) by control point and also aggregated for supplier outbound control point plus related customer inbound control point.

Statistics should be provided to analize the relevance of the alert parameters in the

5 Responsibilities

In general the roles and responsibilities of the partners remain unchanged - however there will be additional inter-company activities. It is the responsibility of every company to negotiate and agree on a supplement to the general Logistics Agreements that describes all responsibilities on a detailed level, e.g.:

- The participants define objectives and KPIs and keep track of those metrics.
- Every partner is responsible to keep its data updated and ensures a defined data accuracy. A KPI for demand forecast quality and the definition of an alert for missing/late data update is will be of great help.
- Parameters (lead-times, alert thresholds, etc.) should be *mutually agreed* with the concerned customer or supplier.
- Clear definition of who is taking care of what kind of alerts (predefined workflow).
- Commitment of all participants to transparency and openness.
- Definition of responsibility for system performance and helpdesk.
- In case of ambiguous information receiver of information is obliged to clarify the situation with sender.
- Every participant defines a central contact person for general coordination purposes.
- If required, the participants agree on a central/global person coordinating the complete supply network.

Data quality is one of the key success factors of SCMo. For this reason plausibility checks and rigorous detection and elimination of the root causes of false alerts are fundamental and needs to be established in the work procedures.

6 Data Flow and Description

Fig.17 illustrates the information flow in a decentralized architecture enabling interoperability between SCMo applications from different software vendors.

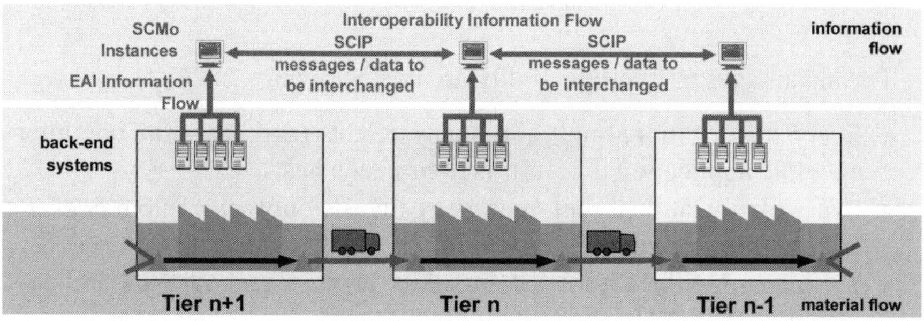

Fig.17: Basic information flow with decentralized architecture

SCMo deals with 3 *categories* of data:

- *SCMo-internal data* is update generally via browser interface. It includes organizational unit parameters, control point parameters, transportation link parameters and production link parameters.
- For the *information flow between backend systems* and the *respective SCMo Instance* ("EAI information flow", vertical) company individual integration concepts required. Basic data sets are: demand information, inventory information, shipments, goods received, work in process inventory (optional).
- The *information flow between SCMo instances* ("Interoperability information flow", "horizontal information flow") is key for interoperability. Basic data sets are: demand communication, inventory visibility, alert visibility, and supply chain model information.

The tier n supplier in Fig.17 may select and install *one interoperable* SCMo application. It may choose a lean low-cost solution or a solution with advanced backend integration features, and user management functionality according to the own needs and preferences.

This SCMo application ("instance") is linked to the company's back end systems (legacy, ERP, warehouse management system, etc.), which remain the leading systems for inventory information and parameters.

After a status change in the back end system (e.g. a shipment is effected and therefore the inventory level of part number xyz in the outbound warehouse drops from 200 to 100 units) the own company's SCMo instance is updated via the (vertical) *EAI information flow*.

In order to inform also the relevant business partners about the new status the interoperability information flow is used to update (synchronize) the adjacent SCMo instances. The SCMo instances of the different organizational units communicate via the Internet, ENX, ANX or other appropriate networks.

The advantages of interoperability are the following:

- Every organizational unit / company selects (and pays for) one interoperable application that fulfills their needs best.
- Every organizational unit / company invests only *once* into a *backend integration* and will be *linked to all business partners*.
- Seamless, highly automated business process and fast and reliable data flow with all business partners.
- Full B2B functionality.

7 Conclusion

SCMo is a mature concept and has already shown its value for mastering critical supply networks. SCMo is an add on to existing concepts. That means, there are only weak interdependencies to existing processes and systems and a re-engineering is not necessary to use the concept. However, the improvements due to SCMo will create room for significant optimization and streamlining.

A nice *analogy* may explain the value of the new concept. SCMo equals the headlamps of a car: During the day or on an illuminated road we do not need headlamps for driving. That means we do not need SCMo in logistics for uncritical supply networks. Like somebody who never had headlamps and never drove in the dark, we will not even miss SCMo. However, on a dark road, we need to reduce our speed significantly and sometimes we will even get off the road. That means in critical supply networks we built up high inventories (slow speed) and sometimes we will have a stock-out. By using headlamps we get the visibility that allows us to drive save and fast in the night. And by using SCMo we get the visibility that allows us to ensure safe supply and to operate with low inventory levels in critical networks. If we used to drive without headlamps in the dark and we get now the chance to try a car with headlamps, we would never take a car without headlamps any more! The same could happen for SCMo.

New Networks: Features and Tasks

Martin Schottenloher

ewf europe and Mathematisches Institut, Ludwig-Maximilians-Universität München

I describe briefly the new networks in the global economy. I concentrate in the characteristics in comparison with the usual networks. I don't explain why the industry is interested in these new networks nor do I discuss the competitive advantages.

Features

Networking is not new in the economy. Many firms and organisations successfully co-operate in partnerships, alliances and networks in order to achieve competitive advantages.

Manuel Castells [1] describes five major kinds of networks in the global network economy:

1. *Supplier networks*, in which firms subcontract for a range of inputs, from design operations to the manufacturing of component parts
2. *Producer networks*, made up of companies that pool their production facilities, financial resources, and human resources in order to expand their portfolios of goods and services, broaden geographic markets, and reduce up-front risk costs
3. *Customer networks*, which link together manufacturers, distributors, marketing channels, value-added resellers, and end users
4. *Standard coalitions*, which bring together as many firms as possible in a given field with the purpose of binding them to the technical standards established by an industry leader
5. *Technology co-operation networks*, which allow firms to share valuable knowledge and expertise in the research and development of product lines.

The current competitive challenges, however, seem to require an even stronger form of co-operation. The recent contributions to the subject of networking and collaboration lead to a vision of new networks that cannot be restricted to one of the above five categories alone. Instead, these future networks incorporate all the five types and they have further important properties. As a consequence these networks are very complex.

The essential ingredient beyond the attributes described in the five categories above seems to be that such networks establish a considerably closer integration and tighter connection of the participating firms or organisations: the network has to function almost like a single company or at least like a joint venture under a new corporate identity, – in some respect like a virtual corporation.

This higher degree of integration manifests itself through

- the fact that all five categories are incorporated.
- the excellence of the network. Excellence is necessary because of the new challenges. High-level expertise is required concerning the objectives of the projects and at the same time in the high competence concerning the ability of collaborating in a network. Because of this demand for excellence I call such a new network a "Network of Excellence" (NoE).
- the fact that challenges require not only optimised co-operation and the sharing of knowledge and information but also the sharing of risk and benefit, the sharing of resources like human resources or research facilities as well as joint establishment of financing.
- the deeper significance of essentially all relevant attributes of a usual network like e.g. communications, responsibility, trust, etc.

The forms and properties of these networks of excellence have, as yet, not been formulated. Several suggestions have been made. They all emphasize closer relationship and a greater importance of properties such as

- communications,
- responsibility
- true partnership,
- trust and confidence,
- best practice,
- sharing of risks and benefits,
- knowledge and learning,

among others.

For Example, Communications

Let me explain the deeper involvement of the basic features of a network of excellence by looking at the increasing importance of communications in such networks.

From the point of view of information theory communication may be understood as the mere transmission of messages. As such there are interesting aspects and challenges which have already been investigated by Norbert Wiener and Claude Shannon, more than fifty years ago.

This point of view of specialists misses the relevance of communications for culture. In the words of Clifford Geertz [2] culture is defined by "the webs of significance" which human beings spin around themselves, and communications – language, art, written text, music, film, software – are the tools, which we use to interpret, reproduce, develop, maintain, and transform these webs of meaning. Communication may therefore be regarded as generating social meaning through the transmission of content in different ways.

Similarly, for the new networks of excellence communications are much more than mere transmission of messages, because communications become the basis of collaboration and serve to bind the network together. It is no accident that communication and community have the same root. Indeed, the networks of the future have to be a kind of communities. Communications are the tools, which create the essence and meaning of the communities in addition to managing the exchange of information.

Communities exist through common meaning and through common forms of communications. A basic characteristic of communities is to share the means by which the community is defined. In the case of networks of excellence the sharing of resources, knowledge, information, risks, benefits and opportunities is an essential property. Only if sharing of the basic ingredients works, the network itself will be successful.

Access

The willingness to share resources implies that all members are provided with the possibility of access.

Networks of Excellence can be understood as a modification from simple partnering or networking with the aim of better selling or producing to the much more tightly connected networks or communities described above. In these NoEs access is provided to resources of all kind (experts, knowledge, new results, information, opportunities, connections, risks, ideas, ...). Stressing access in this context is in accordance with the book of Jeremy Rifkin [3] who describes how ownership in the economy tends to become an anachronism while access is gaining importance.

The basic goal of the NoE would therefore be to create a platform providing access to all sorts of means. It is clear from this description that improved access would be restricted to the participants of the NoE excluding those who are not in the network. In a competitive arena this is exactly what the participants are expecting and demanding.

Implications

Networks of Excellence will change the participating companies. Well-established boundaries between companies have to disappear and the usual hierarchies will not work the same way: People engaged in a network project will feel responsible for this networked project in a stronger form than they feel responsible for the matters of their company. In particular, in the network project they have to work for a team which is in general directed by a project manager of another firm.

In order that such problems can be settled and the network really become successful, a new form of co-operation has to be created on all levels, a new set of rules in a group of companies working together. These forms are not yet known. There exist ideas, however.

These forms will change the organisation of the companies. It will lead to a new form of responsibility of the staff, a new hierarchy, and it implies training of the personnel. It may even imply a repositioning of the firm..

In particular, communications become more and more important. As is well known, in a company itself corporate communication is indispensable in order to concentrate on the core competencies and to survive in a competitive environment. In a supply chain alliance, too, a common identity has to be defined and communicated if the supply chain wants to be successful, as described by Rainer Kurek and Sabine Schindler [4]. Similarly, common goals and a clearly formulated strategy belong to the preconditions for building a successful NoE.

Another basic feature of such NoEs is that partners who co-operate in one network may be competitors in another environment, alliance or network. They will have to master these different roles and to be able to manage the switch from one role to the other. Co-opetition is the word often used in this context.

The new networks are planned to be equal-partner networks not dominated by one of the participating firms. This form of co-operation is difficult to manage since it is in some sense in conflict with the close relationship and tight integration of an NoE described above.

It is apparent from what I have described that a company has to fulfil quite a few of preconditions in order to become a reliable and valuable member of an NoE. To make a company prepared and qualified for such a venture is not an easy task, and this involves not only the company as a whole but more particularly it requires qualification of personnel.

Tasks

How do we find these new forms of collaboration? Who can set the new rules? Clearly the development of such forms and rules cannot be invented or constructed in theory, they have to be found in close co-operation with companies interested in NoEs.

A reasonable first step seems to be to build an NoE as a project. Under this assumption, there are two ways to proceed within the framework of NoEs:

- Learning by doing in a carefully selected NoE project i.e. the new forms and rules of collaboration will be developed together with the objective of the project.
- As a project of a NoE with the objective of developing and evaluating such new forms of collaboration. Hence, the subject of this project will be the rules of NoEs itself.

How will NoEs be organised? A reasonable possibility is a consortium of smaller networks. Fig.1 illustrates this.

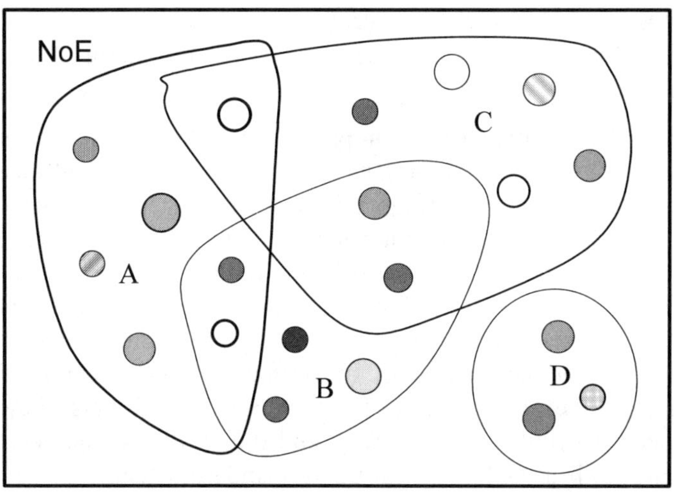

Fig.1: Projects A, B, C, D with their constituent firms (circles) are NoEs in themselves.

For different projects in the automotive industry, for example, groups of companies will come together for each project (possibly with varying participants) in a consortium. The consortium provides a framework for regulating responsibilities, project management, intellectual property, dissemination, etc. Several consortia (i.e. NoEs) may thus be established, one for each project. An essential aspect is that these small NoEs are willing to share their results with other project oriented NoEs. In this way they become a bigger NoE. And, they can be viewed as forming a consortium of consortia. The totality of consorts in the consortia can be regarded as constituting the NoE.

In order to manage the overall programme of such a network and to distribute the relevant information and results in a fair manner a neutral entity and programme leadership is necessary. This can be done by ewf europe.

Among others, the tasks of such a neutral instance for the NoE are

- formulation and management of the programme, while respecting in particular the goals and strategies of the participants of the NoE,
- establishment of a communication platform,
- facilitation of the conflicting interests of the participating firms,
- provision of transparency,
- dissemination of the results, so that the different groups are able to benefit from the results of all the projects,
- scientific monitoring and coaching / auditing,
- provision of examples of the relevant contracts,
- establishment of a programme providing long-term strategies and goals.

Networks tend to disintegrate, in particular, in the case of an equal-partner network. Therefore the functions and tasks of a neutral instance have to be carried out through an efficient network management. The NoEs depend on the establishment, maintenance and strengthening of the relationships in the hope of benefits and profits in the future.

The global economy of the future will undoubtedly see a growth in the number of networks and a change in how they are organised and managed. The NoEs will reduce uncertainty in the face of world economic activity resulting from the globalisation of technology and markets, they will increase speed, flexibility and capacity and they will provide access to resources and skills as well as to information and knowledge.

References

1 Castells M (2000) The Information Age: Economy, Society and Culture, vol 1, 2nd ed.,The Rise of the Network Society. Cambridge, MA, Blackwell

2 Geertz C (1973) Thick description towards an interpretation theory of culture. In: Interpretation of Cultures. Selected Essays. New York, Basic Book, 3-30.

3 Rifkin J (2000) The age of access. London, Penguin Books

4 Kurek R, Schindler S (2002) Gewinner von morgen handeln heute. Stuttgart, Deutsche Verlagsanstalt (DVA)

Networks of Excellence: Ewf Europe's Vision

ewf Executive Board

Herbert Köpplinger, Franz Obinger, Peter Sachsenmeier, Sabine Schindler, Martin Schottenloher

Ewf Europe's Motivation and Values

Ewf as an association cooperates Europe-wide in order to offer holistic methods and solutions for innovative processes in the world of work.

Our concepts and ideas take into account and reflect current trends. We intend to contribute to lasting improvements among our members and partners.

The intended results of our actions have societal significance, they contribute to growth and lead to economic success.

In order to reach the goals we have set ourselves we work in open and honest dialogue. The quality and integrity of our relationships are of importance to us.

Our Vision for Networks of Excellence

The intended Network of Automotive Excellence (NoAE) is our most advanced NoE project. It exemplifies our vision.

The NoAE is a network for companies in the automotive industries which intends to improve the competitive position of those involved. Potentially, we address all the companies in the supply chain.

Competitive abilities are improved through cost savings, quality improvements and the active meeting of new challenges.

In the framework of the NoAE, cost reductions and quality improvements will be achieved through new forms of partnering and collaboration.

Therefore, one of the first tasks of the NoAE will be to address work on new organizational forms and new rules for trusted collaboration.

Ewf is the initiator of the NoAE. It will lead the NoAE in the role of a neutral facilitator, and will deliver a corresponding programme of work. By the time of our biannual conference scheduled to take place again in

2004, we hope to have implemented the NoAE and to be able to showcase our first successes.

Goals

The basis for successful collaboration beyond the confines of companies and industrial segments is the presentation and work on new forms of co-operation.

These forms of cooperation will show several values and behaviours as being fundamental and will incorporate them into a new rule book; such as

- trust
- neutrality
- ability and preparedness to cooperate
- sharing of risks and benefits.

These cooperative forms in the network are the basis for access to resources (information, technologies, innovations,…) which can be provided better, faster and cheaper through a joint effort.

Excellence is a goal.

Strategy

The concept of the NoAE must be tried out in projects. The goals of the NoAE will be achieved through successful projects, or at least, the groundwork will be prepared for these.

Organisationally, the NoAE is the totality of a number of projects which will be conducted by a changing set of companies under the programme leadership of ewf.

We intend to deliver clearly defined projects, fast.

We will work out clear guidelines which will assure the dissemination of results to all those involved in the NoAE.

www.ewf-europe.com

ewf europe

Ewf Association in Switzerland

Jürg A Masson

President ewf association Schweiz; Chairman of the Board, IMAG Information Management AG, Schaffhausen, Switzerland

A lively Engineering Workflow Association exists in Switzerland, founded in January 2000. Our aim is the organisation of a systematic exchange among the various roles of our members, such as users, enablers, integrators and innovators, in order to provide an effective ewf engineering solution.

The association is keen on providing added value to its members, and we do this as follows:

- Exchange of experiences
    - Best practices
    - Lessons learned
- Knowledge Updates
    - Trend scouting
    - Access to state of the art knowledge
- Practical Support
    - Project assurance
    - Holistic view and support
- Networking

Fig. 1: ewf association added value

We use a range of instruments and so far, we have been able to provide a fairly impressive range of activities. These include working circles on knowledge management; retreats at the abbey of Bigorio (annual event); exchange forums, e.g. *From 2D to 3D construction*; seminars and workshops, e.g. on *Knowledge Management in Industry*; a substantial number of presentations to companies, groups and individuals; public lectures, such as a series of lectures in conjunction with the Zurich technical university; and inclusion of engineering workflow in university courses, e.g. a study course in information science at Lucerne.

We have about 40 members; these represent 400 interested experts and firms, and these in turn are associated with 4000 other associations and industrial bodies in Switzerland. Our network in itself is fairly substantial.

We have an active executive committee, a permanent discussion circle on knowledge management, and a distinguished advisory panel. We work together with *TR Technische Rundschau* as the official publication of our association, and also with a trade publication, *SMM Schweizer Maschinenmarkt*, as channels of dissemination.

We are active in several Swiss national industrial initiatives, such as SAP Schweizer Automatik Pool, CCSO Centre CIM des Suisse Occidentale, and STEPS Swiss Technology Partnership.

Our funds come from both membership contributions (including some sponsoring), and earnings from our activities as an association. Our earning are derived from projects, publications, events, and training.

We publish a regular electronic newsletter, *ewf association news*, which can be obtained through our website.

www.ewf.ch

ENGINEERING WORKFLOW ASSOCIATION
Wiesenbachstrasse 7
CH-8542 Wiesendangen
Telefon +41 52 337 37 33
Telefax +41 52 337 37 58

Authors

Becker, Wilhelm

Professional Career:

1970 – 1974:	MTU Munich
1974 – 1976:	Metro Düsseldorf
1976 – 2000	BMW Münich, Purchasing & Logistics
2000 onwards	BMW Munich, Product Line Responsibility for Smaller & Niche Models

Bugow, Rainer

Dr.-Ing. Rainer Bugow is Chief Technology Officer at PROSTEP AG in Darmstadt (Germany). In addition to setting the course for PROSTEP's technology, he also directs the company operations and product development.

Rainer Bugow received his diploma in Mechanical Engineering from the Technical University of Clausthal (Germany). From 1988 to 1994 he worked as a research associate at the Institute for Mechanical Engineering of TU Clausthal. He received his Ph.D. on the "Provision of Standard Parts Libraries in the design process" in 1995. 1995 he joined PROSTEP as Project Manager. After holding management positions in Product Development and Product Management he now serves as Vice President in the PROSTEP management board, since 2000.

PROSTEP AG, Dolivostraße 11, D-64293 Darmstadt

Bullinger, Hans-Jörg

Prof. Dr.-Ing. habil. Prof. e.h. Dr. h. c. Hans-Jörg Bullinger was born in Stuttgart (Germany) on April 13th 1944. He began his career as a worker for the Daimler-Benz company in Stuttgart, after which he read for a degree at the University of Stuttgart, graduating with a Master's degree and Ph.D. in Manufacturing.

At the Fraunhofer-Institute of Production of Technology and Automation in Stuttgart, Dr. Bullinger was responsible for many applied research projects in the field of design and manufacturing, with the goal of increasing the productivity of computer based systems. These projects were carried out together with German companies, such as Robert Bosch, Siemens, VW, Daimler Benz. He has rich knowledge regarding the international situation of applied informatics.

After lecturing at the University of Hagen, Dr. Bullinger was appointed a full-time lecturer at the University of Stuttgart. Apart from his role as chairman of the University, Hans-Jörg Bullinger is also the head of the Institute for Human Factors and Technology Management (IAT) and the Fraunhofer-Institute for Industrial Engineering (IAO). The institutes are renowned for the carrying out of projects with both German and foreign countries.

On October 1, 2002 he became President of the Fraunhofer-Society, the largest Institution for applied research and development in Europe.

Carosio, Stefano

D'Appolonia S.p.A., Via San Nazaro, 19 , I 16145 Genova
stefano.carosio@dappolonia.it

Döbrich, Udo

SIEMENS, Automation and Drives, Advanced Technologies & Standards, Advanced Control and Field Technology

Siemens AG, A&D ATS 3, Östliche Rheinbrückenstrasse 50, D-76187 Karlsruhe

Dokter, Karl

Since 1990, Karl Dokter has a leading function within the EDAG Group. After studying Mechanical Engineering he gained his first experience as a Design Engineer on a Mercedes project in Graz. Projects included design of body and trim parts for BMW and MAN in Munich. Implementation projects for Volkswagen lead him to Mexico, where he gained his first foreign experience.

1990 he took over the subsidiary of EDAG in the UK as Managing Director. In 1993 he became Group Director for Product Development worldwide.

In his new function as Director Key Customer Relation the focus changed to the relationship management of two key customers in Germany and the development of the Asian market for the EDAG Group.

Filos, Erastos

Dr Erastos Filos co-ordinates R&D activities related to "New Methods of Work and Electronic Commerce" within the European Commission's Programme Information Society Technologies (IST). He also follows up several R&D projects funded by the Commission. Before joining the European Research Programmes in 1993 he was responsible for the electronics development team at Bosch Telecom in Heidelberg, was project manager at Perkin-Elmer Corp. in Überlingen and worked as research assistant at the University of Constance in Germany.

European Commission, Directorate-General Information Society, Brussels
Tel +32 2 296 24 92, Fax +32 2 299 28 65
erastos.filos@cec.eu.int

Fischer, Dietmar R.

Dietmar R. Fischer was born on 19th February, 1959, in Germany. After studying Aerospace & Aeronautical Engineering at the University of Stuttgart (1987) he took a PhD in mechanical engineering. Since 1984 he works for the Fraunhofer Society and for the last 6 years he was a member of the managing board. In 2000 he founded AGILeVIA GmbH. Dr. Dietmar Fischer is the managing director of AGILeVIA Business – a spin-off of the Fraunhofer institute for industrial engineering. He is also responsible for establishing the business unit consulting within AGILeVIA.

Gebauer, Gunter

Born in 1944. studied in Kiel, Mainz und Berlin (FU and TU). PhD in 1969 on the language philosophy of Wittgenstein, higher doctorate 1975, in a theory of understanding. Gunter Gebauer is Professor at the Institute of Philosophy at the Freie Universität in Berlin.

Works on historical anthropology, social philosophy, and the sociology of sports. Newer work concentrates on Nietzsche, Wittgenstein, Foucault und Bourdieu.

Heidel, Roland

SIEMENS, Automation and Drives, Advanced Technologies & Standards, Advanced Control and Field Technology

Siemens AG, A&D ATS 3, Östliche Rheinbrückenstrasse 50, D-76187 Karlsruhe
Tel. +49 721 595 4632, Fax: +49 721 595 6728
Roland.Heidel@siemens.com

Höppe, Alexander

Alexander Höppe is a Senior Consultant with IBM Business Consulting Services. Within the high tech and capital goods industries, his specialist fields are supply chain and product lifecycle management. In these areas, he has carried out IT and process consulting as project lead. Furthermore, he has held key functions in two ERP implementation projects, focusing on project management and controlling modules. Prior to working as a business consultant, he gained 4 ½ years of operational project experience in a major power plant and environmental technology company, for which he ultimately held responsibility as project manager. Alexander Höppe graduated with adegree in mechanical engineering from the Ruhr Universität in Bochum (Germany) and has completed an MBA at the Heriot-Watt University in Edinburgh (UK).

Jung, Hans H

Dr. Hans-Hermann Jung ist the Vice President responsible for Customer Relationship Management at Nexolab GmbH, the management consulting subsidiary of the BMW Group. From 1998 to 2001 Hans Jung was responsible for Customer Relationship Management for the Mercedes-Benz brand at DaimlerChrysler in London, developing a European strategy. Precisely, DaimlerChrysler, Hans Jung was a partner at ISUMA Consulting, responsible for marketing projects in various industrial sectors.

Kaplaner, Klaus

Dr Klaus Kaplaner is CEO of Nexolab, a BMW Group Company. Nexolab is a management consulting firm focused on the automotive and manufacturing industry. Mr. Kaplaner obtained his doctorate in business administration at the Ludwigs-Maximilians-Universität in Munich. Before joining Nexolab he was partner of an international management consulting firm.

Nexolab GmbH, Zamdorfer Straße 100, D-81677 München
Phone +49 (0)89-99399-162, Fax +49 (0)89-99399-401
Klaus.Kaplaner@nexolab.com

Köpf, Peter

Born on December 14, 1944, Geislingen/Steige, Germany
Professional Career:
1963 - 1970: University of Karlsruhe
- Studies in mechanical engineering
- Diplom Engineer
1970 - 1978: Mercedes Benz, Stuttgart
Advanced engineering for automatic transmissions
Since 1978: ZF Friedrichshafen AG
Current position: Vice President Corporate Research and Development

Köpplinger, Herbert

Herbert Köpplinger is President of ewf europe. In 1998 he was a founder of AK ewf, the precursor of ewf europe.

From 1995 to 1999 he was CEO of SIFRAME Software Technologies GmbH, based in Munich, Germany. Before SIFRAME Herbert Köpplinger was Director Concurrent Engineering at Siemens-Nixdorf with worldwide responsibilities and Marketing Manager Automotive at Siemens AG.

herbert.koepplinger@ewf-europe.com

Kühl, Stefan

Born in 1966, studied sociology and history at the universities of Bielefeld, John Hopkins, Paris-X-Nanterre and Oxford. PhD in sociology at Munich university, also higher doctorate. Currently on the academic staff at the Institute of Sociology, Munich. Works as an organisational consultant to various international firms.

Areas of interest include theory of society, organizational sociology, sociology of technology, industrial sociology, sociology of professions, history of science.

Kurek, Rainer

Since October 2000 Rainer Kurek has been responsible for IVM AUTOMOTIVE´s Sales Department in Munich.

After completing a degree in Mechanical Engineering and of a dissertation based on his work at AUDI in Ingolstadt, he joined the SEMPREX/ VENTREX Valve Factory in 1994. There, he was responsible for Product Development. Later in 1996 he took on management of the development site in Germany. In the spring of 1998, as Head of the Body-in-White Department at IVM AUTOMOTIVE he directed in Ingolstadt, a project for the design of a saloon car. As Head of the Automotive Section at the renowned Management Zentrum St. Gallen he developed the seminar, *»Winning Strategies in the Automotive Supply Chain«* which he taught himself and which became the starting point for his book, *»Gewinner von morgen handeln heute«* (Tomorrow's Winners Act Today).

Since January 2001 Rainer Kurek has been a member of the board at the »Industrieholding-Beteiligungs-AG« in Vienna, Austria. In January 2002 he founded AUTOMOTIVE MANAGEMENT CONSULTING GmbH in Munich.

Masson, Jürg A.

President ewf association Schweiz; Chairman of the Board, IMAG Information Management AG, Schaffhausen, Switzerland.
Juerg.Masson@ewf.ch
Juerg_Masson@imag-web.com

Merle, Oliver

is 38 years old and lives in Reutlingen, Germany. He studied at the Technical University in Darmstadt, Germany and graduated in Electrical Engineering and Management. In 1991 he joined the Bosch Group where he had first assignments in several function of the Body Electronics Division, the International Purchasing Liasion Office in Singapore the Brazilian Organisation and in the headquarter. In 1995 he became Head of Purchasing in Bosch's Lighting Division and from 1999 Vice President Purchasing of the Automotive Lighting Group. Since August 2000 he is Project Director in the Bosch's Corporate E-Business department with responsility for Purchasing and Logistics. Since March 2001 he is in addition project leader of the Odette Supply Chain Management Group.

Robert Bosch GmbH, Corporate E-Business (ECO), Fabrikstr. 42, 73207 Plochingen
Phone: +49 (0)7153/666-625, Fax: +49 (0)7153/666-80 625
Oliver.Merle@de.bosch.com

Mund, Peter

BMEnet GmbH

Mundt, Hans Erich

Date/Place of Birth: 14.09.1944 / Euskirchen (Germany)
Maritial Status: Married / one Daughter
Education: Engineer (grad.), Branch: Aeronautical Engineering, Graduation in 1968 at the Technical College Aachen (Germany)
Professional Experience in Management Positions in:

- Aircraft Design and Production in the Sectors
- Airframe Design
- Quality Assurance
- Production
- Project Management
- Procurement /Purchasing
- International Projects

Professional Background:

07/00 – present	EADS Deutschland GmbH, Munich Senior Vice President Corporate Sourcing
10/96 - 06/00	DaimlerChrysler Aerospace Airbus GmbH Hamburg, Vice President Procurement
05/92 - 09/96	Dornier GmbH, Oberpfaffenhofen, Vice President Procurement
01/87 - 04/92	Dornier GmbH, Oberpfaffenhofen, Vice President Civil Aircraft
01/84 - 12/86	Dornier GmbH, Oberpfaffenhofen, Vice President Project Management (Alpha-Jet, Airbus, Ariane, IA 63, Do 228, Do 228 licence production in India)
01/80 - 12/83	Dornier GmbH, Oberpfaffenhofen, Project Manager Do 228
10/72 - 12/79	Dornier GmbH, Oberpfaffenhofen, Director Quality Assurance
09/68 - 09/72	Dornier GmbH, Friedrichshafen, Design Engineer Airframe

Nevey, Steve

As Computer Aided Engineering Manager, Steve Nevey is responsible for specifying, procuring, and maintaining CAE hardware and software at Jaguar Racing. He is located in Jaguar Racing's Design Office at the team's design and manufacturing facility in Milton Keynes, England.

Steve joined Jaguar Racing (then Stewart Grand Prix) as IT Manager in 1996, when the team was first established.

Steve started his career in the shipbuilding industry as a Ship Designer, working mainly on submarine structures. His previous Formula 1 experience was with the Footwork (now Arrows) team, where he worked as a Design Engineer for three years. Immediately prior to joining Stewart Grand Prix, Steve was a Senior Teacher with the University of Warwick, where his responsibilities included teaching Design and Design Management at post-graduate degree level, and delivering consultancy to engineering companies throughout the world, notably in India, Thailand, Singapore and Malaysia.

Otto, Boris

Dr. Boris Otto holds a degree in industrial engineering and management from the Technical University of Hamburg and a PhD degree in mechanical engineering from the University of Stuttgart. After finishing his studies, he was a consultant at PricewaterhouseCoopers with a strong focus on the optimization of industrial business processes.

In May 2002, he joined Fraunhofer IAO as a senior researcher. His main areas of work are inter-company business processes, e-business integration and business models for e-business. Since 2002 Dr. Boris Otto is head of the e-business integration unit at Fraunhofer IAO with seven scientific researchers. He was project manager for various research projects and industry assignments. Moreover, he plays an active role in the standardization are for e-business, e.g. in the CEN/ISSS e-Commerce workshop. Dr. Boris Otto is a certified SAP consultant for the MM module. He speaks German and English fluently and has a basic knowledge in Portuguese, Italian, and French.

Péro, Hervé

Since 1999, Hervé Péro has the responsibility of the unit "Innovative Products, Processes and Organisation" within the EC, DG research, and assists the director in preparing the next EC research programme on industrial technologies. Before he worked before as advisor to the director in charge of "Industrial & Materials Technologies", "Standards, Measurements & Testing" and "Steel Research" programmes (1993-98), and as Administrator in the BRITE & BRITE/EURAM programmes (1986-1993). As French engineer, he had working experience within the industry, from 1977 to 1986 (e.g. Head of department in Vallourec, Large Welded Pipelines Division), and within university, in South America (Technological Institute of Caracas, 1973-77).

Plass, Ludolf

Chief Technology Officer Lurgi AG, Vice President R&D /Technology /Project Execution Management Lurgi AG, Vice President Corporate Development Lurgi AG
Age: 58 years

Studied Mechanical Engineering at Darmstadt Technical University. Doctor's Degree: At Chemical / Mechanical Process Engineering Division of the University of Erlangen Nuremberg

33 years' employment with Lurgi in the following functions: Director R&D at Lurgi Chemie und Hüttentechnik GmbH; Head of Coal and Energy Technology Division of Lurgi Kohle- und Mineralöltechnik GmbH; Managing Director Lurgi Energie und Umwelttechnik GmbH; Chairman of the Board of Management Lurgi Energie und Umwelttechnik GmbH

Longstanding board member in numerous German and international subsidiaries of Lurgi AG

Since 1996 Chief Technology Officer at Lurgi AG responsible for:

- R&D and Technology Management at the Group
- Optimization of Business Processes
- Corporate Development / Portfolio Management

Activities outside Lurgi AG: Member of the Board of Dechema and of DVCV: Member of the Advisory Council of Fraunhofer IPK Berlin; Member of the Advisory Council of Schaefer Kalk KG, Diez/Lahn.

Reinheimer, Thorsten

Dr Thorsten Reinheimer is the Director Consulting of MOSAIC SOFTWARE AG. Strategic development of products and solutions and research are part of his function as well as consulting in the field of E-Commerce /EDI, from process level to the solution architecture. He started working for MOSAIC SOFTWARE AG in 1997 and has been part of the Management Board since 1998.

Dr Thorsten Reinheimer studied physics and graduated at the Max-Planck-Institute for Radio-Astronomy in Bonn (applied optic and digital image processing). He continued working on fundamental research for another three years, before accepting a job in industry.

He was born in 1962, and he lives near Bonn.

MOSAIC SOFTWARE AG, Feldstr. 8, D-53340 Meckenheim

Phone: +49(092225/882-0, Fax: +49(092225/882-201
E-mail: info@mosaic-ag.com

Sachsenmeier, Peter

Peter Sachsenmeier, Professor of Information Management, is ewf europe´s Vice President in charge of the Network for Excellence idea. He is also joint CEO of IMAG Information Management AG, a consulting firm based in Schaffhausen, Switzerland.

For many years, Peter Sachsenmeier has advised international companies and organizations on their strategies and IT governance, information architectures and knowledge management. He has particular expertise in the planning and implementation of very large international information and communications projects.

IMAG Information Management AG, Blumenaustr. 28, CH-8200 Schaffhausen
Peter.Sachsenmeier@ewf-europe.com,
Peter_Sachsenmeier@imag-web.com

Schäfer, Anna-Maria

Studies, Stuttgart: 1985-89, Diploma Degree „Information and Documentation /Information Science"
Professional Experience:

1989-1991	Frankfurter Allgemeine Zeitung: Information Specialist
1991-2000	McKinsey & Company: Automotive & Assembly Practice, Research Analyst
2000-	LBBW Landesbank Baden-Württemberg, Corporate Finance, Focus Industries Automotive and Machinery, Industry Analyst

Scherer, Eric

Dr. Eric Scherer is managing director of intelligent systems solutions (i2s) GmbH, Zurich, Switzerland.

Dr. Scherer studied mechanical and industrial engineering at the Technical Universities of Stuttgart and Aachen (both Germany) and at Rutgers University, USA. He received his Ph.D. in industrial management from the Swiss Federal Institute of Technology (ETH) in Zurich. Before joining i2s he worked in the information systems and services unit of the GM/Opel International Development Center in Rüsselsheim, Germany.

He is member of the board of the engineering workflow (ewf) association in Switzerland.

intelligent systems solutions (i2s) GmbH, Stampfenbachstr. 159, CH 8006 Zürich, Switzerland

Tel. +41-1-360.51.30, Fax +41-1-360.51.32, e-mail: scherer@i2s-consulting.com

Schmidt, Martin

Diplom-Informatiker, born in 1963.

From 1990 to 1994 Research Assistant at the Institute of Computer Applications in Planning and Engineering at the University of Karlsruhe; 1995 to 1996 Research Assistant at the Research and Technology Department of Daimler-Chrysler AG. Since 1997, responsible for the implementation of PDM solutions at debis Systemhaus. Since 2000 Director Business Center Engineering Process and Data Management Solutions at T-Systems.

Erich-Herion Str. 13, D-70736 Fellbach

Scholz, Alexander

Education

1989 - 1995	Studies of Business Administration, University of Regensburg; Majors: Organisational Development, Human Resource Management, Economic Policy and Econometrics
1991 - 1992	University College of Dublin; Majors: Marketing and International Marketing
1994 – 1995	Université Jean Moulin Lyon III ; Major: International Finance, Portfolio Management and Risk Management

Professional Experience

10/95 - 12/98	Accenture Germany; Business Process and Change Management Consulting, various implementation projects within the financial services and professional services industries
01/99 – 12/00	Oracle Germany, Solution Consulting; e-Business strategy consulting for various customers, focus on electronic marketplaces, e-procurement and CRM
01/01 - 09/01	fashionXchange AG; electronic B2B marketplace for the textile and apparel industry
Since 10/2001	Chief Strategy Officer and Chief Technology Officer; Nexolab GmbH, Manager, Sourcing & Procurement

Schottenloher, Martin

Martin Schottenloher is a member of the executive board of ewf europe, and he is professor of mathematics at the Ludwig-Maximilians-Universität München. His research interests are applications of geometry, e.g. computer graphics, geometric models of quantum field theory, and the latest developments of topological field theory with the possible impact on quantum computing.

Mathematisches Institut, Theresienstr. 39, D 80333 München
Tel.: +49 (0) 179 575 3155
martin.schottenloher@mathematik.uni-muenchen.de,
martin.schottenloher@ewf-europe.com

Spee, Jacques

Jacques Spee, senior consultant at EDS Solutions Consulting, has over ten years of experience in the application and improvement of product development with multi-national multi-discipline teams. He consults management teams and leads transformation programs that focus on time-to-market and innovation for industry clusters, paying attention to both the contents and the participants in the change process. Jacques graduated from the Technical University of Delft in Aerospace Engineering and worked in the aerospace, high-tech and transportation industries. He is author of publications on complex systems development, and process management in product development.

Speiser, Reiner

Reiner Speiser is Vice President at Nexolab GmbH, responsible for the Collaborative Engineering practice. From 1998 to 2001 he was a partner at Median Management Consulting and, prior to this, was a founder and director of ixtel Consult GmbH. His prime involvement has been in strategy and restructuring projects in the automotive industry and the capital equipment and machine building sector.

Reiner.Speiser@nexolab.com

Stehle, Peter

Born: November 17, 1940
Education: Technical studies at the University of Aachen, Germany
Professional experience:

1967 – 1971:	Georg Stetter Maschinenfabrik KG, Memmingen
1971 – 1982:	Brown Boveri + Cie AG, Mannheim; Managing

Management Experience

1982 – 1986:	Diehl, Nürnberg; Chairman of the Board
1986 – 1990:	AEG AG, Frankfurt; Member of the Board
1990 –	Freudenberg & Co

Teifel, Harry

Born in South Africa – 1969; schooling in South Africa
Studied Industrial Engineering, MBA and DBA
Work experience:

- Manufacturing industry: 4 years
- IBM Consulting: 4 years
- DEKRA: 3 years

Current position: Director of DEKRA strategy, dealing with all strategic issues including technology

Wagner, Reinhard

In 1985 Reinhard Wagner started his career in the German Air Force. Since 1995, after studying Electrical Engineering and Business Administration in Germany and the USA, he has been engaged in the field of Engineering Services. In 1996 he took over the subsidiary of FERCHAU Engineering Services in Ulm as Managing Director and left in 2000 for the headquarters of IVM Automotive in Munich, a leading service provider for total vehicle engineering activities. Responsible for human resources and project management he headed all activities for continuous improvement, especially the development of personnel. With this background he recently started as an independent trainer and consultant for human aspects in complex project environments. Since 2001 he has also been lecturer at the University of Applied Sciences in Neu-Ulm.

Wenzek, Hagen

Professional Experience:

02/02 - today	Industry Leader Aerospace & Defense, IBM
01/01 - 01/02	Sr. Consultant e-Strategy & Change, IBM
10/99 - 12/00	Consultant e-procurement, IBM

01/97 - 09/99	Research assistant and consultant for automation technology; doctoral studies, University of Hagen
07/98 - 09/98	PhD Scholarship at the Tokyo Institute of Technology, Japan
08/95 - 12/96	Consultant and research assistant, Research Institute for Operations Research at the University of Aachen
10/88 - 05/95	Electrical engineering studies, Aachen Univ.

Wolff, Roland

Education

1979 – 1983	Dipl. Ing (FH) Chemical Engineering
1987	Industrial Engineering, Columbia University, New York

Career

1983 -1984	Chemical Engineer, Biochemical Research & Development, Degussa AG
1985	Trainee Production/Sourcing/Logistics, Wella AG
1986	Assistant V. P. Operations, Wella Inc. USA
1987- 1988	Manager Central Planning, Wella Inc. USA
1989 - 1991	Vice President Operations, Goldwell Inc. USA
1991 - 1992	Head of Corporate Logistics, Stada AG
1993 – 1996	Distribution & Information Technology Manager, Pfizer Hospital Products Group
1997 - 1999	Managing Consultant, Supply Chain Management Group, Ernst & Young Consulting
2000 - 2001	Head of eSourcing for Central Europe, Cap Gemini Ernst & Young Consulting
2001-	Head of Competence Center Sourcing & Procurement, Nexolab GmbH

Programme of the Conference

Tuesday, October 29th, 2002

8.30 a.m.	**Registration, issuance of conference documents, opening of the exhibition**
9.30 a.m.	**Welcome ewf europe**
9.40 a.m.	**Welcome AGORIA**
9.45 a.m.	**The Future of European Manufacturing** Prof. Dr. Hans-Jörg Bullinger, Fraunhofer Institute for Industrial Engineering
10.20 a.m.	**The European Research Area: A Challenge for Co-opetition** Dr. Hervé Péro, Head of Unit, EU -- DG Research
10.55 a.m.	**Coffee break / visit to the exhibition**
11.55 a.m.	**Plant Engineering in Global Competition** Dr. Ludolf Plass, Chief Technology Officer LURGI AG
12.30 p.m.	**Lunch / visit to the exhibition**
2.00 p.m.	**Parallel running discussion forums** **Forum 1: Secure Engineering Collaboration** **Forum 2: e-Business: The Importance of Standards and e-Marketplaces** **Forum 3: Neutrality, Openness and Co-operation in the Service-industry**
3.30 p.m.	**Coffee break / visit to the exhibition**
4.15 p.m.	**Competition and Collaboration in Formula One: A Special Market** Steve Nevey, Computer Aided Engineering Manager, Jaguar Racing
4.50 p.m.	**Long-Term Cooperation in the European Aerospace Industry** Hans Erich Mundt, Senior Vice President Corporate Sourcing, EADS Deutschland GmbH
5.20 p.m.	**Discussion and end of the first day presentations** Prof. Peter Sachsenmeier, Conference Chairman
7.00 p.m.	**Evening event**

Wednesday, October 30th, 2002

8.30 a.m.	**Changes in Development, Production and Branding Policy** Wilhelm Becker, BMW Group, General Manager Product Line, Medium Sized Vehicles
9.05 a.m.	**The European Automotive Supply Industry at the Crossroads** Dr. Peter Stehle, Managing Director, Freudenberg & Co. KG
9.40 a.m.	**Network of Automotive Excellence – The Future Role of the Tier 1 Supplier Viewed from Development** Peter Köpf, Director Central Research & Development, ZF Friedrichshafen AG
10.15 a.m.	**Coffee break / visit to the exhibition**
11.15 a.m.	**Parallel running discussion forums** **Forum 4: Recognizing, Developing and Implementing Core Competences** **Forum 5: 30% More Efficiency in Product Development** **Forum 6: Aerospace Industry: Small and Medium Sized Enterprises Collaborating on a Virtual Level** **Forum 7: Technology as a Basis for Collaboration**
12.45 p.m.	**Lunch / visit to the exhibition**
1.45 p.m.	**eEurope and the European Research Area – Challenges for Business & Work** Dr. Erastos Filos, European Commission
2.20 p.m.	**Shaping the Future Together: The Network of Excellence** Prof. Peter Sachsenmeier, Vice President ewf europe and IMAG Information Management AG
2.55 p.m.	**Final address and discussion** Herbert Köpplinger, President ewf europe Prof. Peter Sachsenmeier, Conference chairman
3.10 p.m.	**Coffee break / visit to the exhibition**
4.00 p.m.	**End of the 2nd e.e.u.c. 2002**

Participants

Anderl, Prof. Dr. Reiner	TU Darmstadt
Aubel, Thomas	TÜV Automotive GmbH
Auerswald, Christoph	Wiltronic AG
Balk, Axel	Volkswagen AG
Barbagelata, Andrea	T4TECH srl
Becker, Wilhelm	BMW AG
Bellenhaus, Franz	BMW Group
Berchtold, Dr. Gerd	EADS Military Aircraft
Betz, Raymond	Solvay
Brändle, Mathias	SGS TPS GmbH
Brisson, Pierre	European Space Agency
Broughton, Jim	Airbus
Buddrus, Manfred	Hewlett Packard GmbH
Bugow, Dr. Rainer	ProSTEP AG
Bullinger, Prof.Dr. Hans-Jörg	Fraunhofer Institut IAO
Campigli, Silvio	Grado Zero Espace srl
Canvin, John	Odette
Carl, Constantin	Dynamit Nobel
Carosio, Stefano	D'Appolonia SpA.
Chiarini, Paola	AECMA
De Keersmaeker, Dirk	T-Systems Belgium NV/SA
De Vocht, Luc	Agoria
Dietz, Helmut	ITI
Doese, Manfred	Siemens AG
Dokter, Karl	Edag
Eichenauer, Heinrich	Covisint
Escabias, Domingo	Bonné GmbH
Fehr, Thomas	Bonné GmbH
Fellows, Andrew	ewf europe
Filos, Dr. Erastos	European Commission
Fischer, Dr. Dietmar	AGILeVIA GmbH
Fourgo, Arabelle	EUREKA
Frank, Dr. Thorsten	Freudenberg NOK Mechatronics KG
Franzeck, Dr. Jens	EADS Deutschland GmbH
Fremuth, Otto	T-Systems International GmbH

Friedrich, Jürgen	T-Systems Work-Center
Furtmeier, Anton	F+F Computeranwendungen
Gassner, Dr. Oliver	Freudenberg & Co.
Gohmert, Kurt	T-Systems International
Gönnheimer, Christoph	Universität Karlsruhe
Groll, Fried	Ernst&Young AG
Grunau, Dr. Arbogast	INA Schaeffler KG
Grundke, Manfred	Bosch Rexroth AG
Heidel, Roland	Siemens A&D
Helmold, Dirk	TÜV Nord Gruppe
Herold, Sabine	DELO Industrie Klebstoffe
Hermany, Michael	Siemens AG
Hinoul, Martin	Flemish Ministry of Economics
Hocevar, Simona	The Ministry of the Economy
Hödl, Hubert	Magna Steyr AG
Hoffmann, Dr. Stefan	SupplyOn
Homsi, Philippe	Airbus, Toulouse
Huber, Charles	Fachhochschule Aargau
Hurst, Ken	Conquest Business Magazine UK
Ittner, Ariane	T-Systems International
Jacoby, Ralph	Bertrandt AG
Johanson, Björn	ESTIEM
Jonckheer, Frans	Sequoyah International Restructuring
Jung, Dr. Hans	Nexolab GmbH
Kallabis, Dr. Matthias	TMS Produktionssysteme GmbH
Kaplaner, Dr. Klaus	Nexolab GmbH
Kattenstroth, Ulf	Benteler AG
Keijzer, Willem	Nexolab GmbH
Klauser, Gerhard	Wiltronic AG
Klement, Thomas	Wirtschaftsförderung Moselle
Knoors, Frank	Sequoyah International Restructuring
Köpf, Peter	ZF Friedrichshafen AG
Köpplinger, Herbert	ewf europe
Kosean, Werner	Husco International GmbH
Kostwein, Hans	Kostwein GmbH
Krebs, Bernhard	BMW Group
Kugel, Sabine	AGILeVIA GmbH
Kurek, Rainer	AMC
Large, Didier	Hewlett Packard France
Latty, Andrea	AL-Kongressmanagement

Lau, Joachim	CLAAS GmbH
Lay, Meinrad	T-Systems International
Lehmann, Jörn	VDMA e.V.
Loy, Maike	AGILeVIA GmbH
Mähler, Andreas	CLAAS GmbH
Marks, Peter	Robert Bosch GmbH
Mayer, Urban	EADS MATRA DATAVISION GmbH
Müller, Prof. Egon	Fachhochschule Zwickau
Mullins, Sinead	EUREKA
Mund, Peter	BMEnet GmbH
Mundt, Hans Erich	EADS Deutschland GmbH
Nast, Alexander	Mascot Systems GmbH
Nevey, Steve	Jaguar Racing
Niermann, Erich	Eureka Secretariat
Novák, Josef	Education Tech. University Ostrava
Obinger, Prof. Dr. Franz	CEFE, ewf europe, FH Augsburg
Otto, Boris	Fraunhofer IAO
Péro, Hervé	European Commission, Directorate-General for Research
Pfuhl, Heiner	Cooper Tools
Pitschmann, Daniel	LISI Automotive
Plass, Dr. Ludolf	Lurgi AG
Pospischil, Rudolf	Renault Institut GmbH
Reinheimer, Dr. Thorsten	MOSAIC Software AG
Reisbeck, Dr. Tobias	ewf europe
Rempp, Dr. Helmut	Forschungszentrum Karlsruhe
Riepe, Steffen	Bombardier Transportation
Rossi, Gianluca	APRE
Sachsenmeier, Prof. Dr. Peter	IMAG Information Management AG, ewf europe
Schauler, Christian	Nexolab GmbH
Scherer, Dr. Eric	intelligent systems solutions (i2s) GmbH
Schier, André	BMW AG
Schindler, Sabine	Schindler PR, ewf europe
Schmidt, Martin	T-Systems International
Schmitt, Dr. Reinhard	NEXVAL Solutions S.L.
Scholz, Peter	Additive GmbH
Scholz, Alexander	Nexolab GmbH
Schottenloher, Prof. Dr. Martin	ewf europe, Math. Inst. LMU München
Schumacher, Dr. Oliver	Fraunhofer IAO

Siebert, Reiner	DaimlerChrysler AG
Skalicky, Jiri	University of West Bohemia
Spee, Jacques	EDS
Speiser, Reiner	Nexolab GmbH
Spreitzer, Thomas	T-Systems International
Stanek, Jan	Stanek Consulting
Stehle, Dr. Peter	Freudenberg & Co.KG
Sternemann, Dr. Karl-Heinz	Universität Karlsruhe
Stierlen, Dr. Cristof	F+F Computeranwendungen
Teifel, Dr. Harry	DEKRA AG
Truetschel, Thomas	T-Systems International
Vacek, Prof. Jiri	University of West Bohemia
van Acker, Martijn	ESTIEM
van Belle, Hubert	ESTIEM
Wagner, Reinhard	Automotive Management Consulting
Weiss, Dr.	AGENSIS company consultancy
Wendel, Stefan	Alcatel-SEL AG
Wenzek, Dr. Hagen	IBM Global Services
Willi, Bernhard	T-Systems International
Wobido, Thomas	SupplyOn
Wolff, Roland	Nexolab GmbH

Druck: Strauss Offsetdruck, Mörlenbach
Verarbeitung: Schäffer, Grünstadt